THE LIBRARY
ST. MARY'S COLLEGE OF MARYLAND
ST. MARY'S CITY, MARYLAND 20686

Grievance Initiation
and
Resolution

Grievance Initiation and Resolution
A Study in Basic Steel

DAVID A. PEACH
Associate Professor in Business Administration
The University of Western Ontario

E. ROBERT LIVERNASH
Albert J. Weatherhead, Jr., Professor of Business Administration
Harvard University

DIVISION OF RESEARCH
Graduate School of Business Administration
Harvard University
Boston · 1974

© Copyright 1974
By the President and Fellows of Harvard College
All rights reserved

Library of Congress Catalog Card No. 73-93775
ISBN 0-87584-112-0

Faculty research at the Harvard Business School is undertaken with the expectation of publication. In such publication the Faculty member responsible for the research project is also responsible for statements of fact, opinions, and conclusions expressed. Neither the Harvard Business School, its Faculty as a whole, nor the President and Fellows of Harvard College reach conclusions or make recommendations as results of Faculty research.

Distributed by
Harvard University Press
Cambridge, Massachusetts
1974

Printed in the United States of America

Acknowledgments

THE RESEARCH ON WHICH THIS STUDY IS BASED was conducted with the cooperation and financial support of the Committee on Industrial Relations of the American Iron and Steel Institute. We are grateful for the support of the Committee and particularly for the encouragement and advice of two of its chairmen, Leo Teplow and Wayne Brooks. Mr. Teplow's receptivity to the research proposal and his coordinating activities throughout the period of field work were particularly valuable.

The five companies involved in the study have been disguised at all levels of their respective organizations where help and cooperation were instrumental in seeing the research completed. We cannot, therefore, publically acknowledge our debt to the company and union officials. Nonetheless, our appreciation for their efforts—both in seeing the study completed and in the day-to-day task of making collective bargaining work—is very real. We here express our thanks to all of these individuals for all that they have done to help us.

Much of the actual writing and rewriting was made possible by an allocation of funds to the School under the direction of the Division of Research at the Harvard Business

School. We appreciate this support and the efforts of two of the Division's directors, Dean Lawrence E. Fouraker and Professor Richard E. Walton. Miss Hilma Holton provided the final editing and guided the manuscript through to publication.

Finally, we should note that the views expressed are our own, and not necessarily those of either the American Iron and Steel Institute or of the companies and unions studied. We gladly take full and complete responsibility for both the descriptions of events and the conclusions we have drawn from them.

DAVID A. PEACH
E. ROBERT LIVERNASH

March 1974

Table of Contents

Acknowledgments v

1. AN INTRODUCTION TO THE STUDY 1
   The Nature of the Study 2
   The Analytical Framework 13
   Research Methodology 17

2. ENVIRONMENTAL INFLUENCES ON PROBLEMS AND
   THEIR RESOLUTION 23
   Socioeconomic Conditions 25
   Technological Change 32
   Task Organization and Work Environment 42
   Summary 57

3. UNION INFLUENCES ON THE CHALLENGE RATE AND THE
   RESOLUTION PROCESS 61
   Union Leadership 62
   Union Organization 76
   Union Policy 83
   Summary 85

4. MANAGEMENT INFLUENCES ON THE GRIEVANCE RATE
   AND RESOLUTION PROCESS 89
   Management Organization 91

Management Leadership 107
Management Policies 112
Summary 126

5. INTERRELATIONSHIPS AND THEIR IMPLICATIONS FOR MANAGEMENT 131
A Synoptic Review of the Study 133
Low- and High-Rate Syndromes and Interrelations Among the Variables 137
The Implications of the Study for Management 144

Bibliography 153

List of Exhibits

1. Summary of Companies and Departments Studied . 5
2. Grievance Rates in High- and Low-Grievance-Rate Departments Studied 6
3. Breakdown in Grievances 51
4. Summary of Steel Study 132

Grievance Initiation and Resolution

1 An Introduction to the Study

THE GRIEVANCE PROCEDURE CONSTITUTES the central focus of day-to-day union-management activity. We hope that this study can provide some helpful insights into the process of resolving grievances and into the variables that determine differences in the level of grievance activity among departments of a company. The field research, carried out by David A. Peach, was conducted in ten departments of five basic steel companies from August 1967 to October 1968. The entire study was conducted with the cooperation and financial support of the Committee on Industrial Relations of the American Iron and Steel Institute. While directly related to the basic steel industry, the analysis appears applicable to a wide spectrum of union-management situations. In fact, the primary objective of this report is not to give an exhaustive analysis of the problem in the steel industry, but to illustrate and discuss a simple framework for the analysis of departmental differences in grievance activity and in the grievance-resolution process at lower levels of management.

The Nature of the Study

Grievance rates—written grievances per hundred employees per year—differ among departments and plants of the same company, as well as among companies.[1] The study analyzes the reasons for the differences, particularly the differences among departments, and also considers the relative roles of the parties, union and management, in the process of resolving disputes.

Within each of the five companies that voluntarily participated in the study, at least two departments were studied in considerable depth—one with a relatively high grievance rate, and one with a relatively low one. Since the essence of grievance is a charge that the agreement has been violated, differences in grievance rates among departments reflected either differences in the number of charges made or differences in the manner in which complaints, problems, and charges were resolved and recorded by the parties.[2] Grievance rates count and reflect only formal charges reduced to writing. The grievance process, however, involves informal and oral activity as well as formal written grievances.

It was anticipated at the outset that statistical evidence could be developed for the total number of problems, complaints, and grievances—formal and informal in each department. This proved to be impossible. No records were available that could give a reliable indication of the total number of complaints, nor could careful interviewing produce usable data. The obtainable evidence of differences in the rate at which problems arose, of the frequency of com-

---

[1] Sumner H. Slichter, James J. Healey, and E. Robert Livernash, *The Impact of Collective Bargaining on Management* (Washington, D.C.: The Brookings Institution, 1960), p. 698.

[2] Ibid., p. 694.

plaints, and of the resolution process was, loosely speaking, clinical in nature rather than statistical, although formal grievance rates were developed as a starting point. Statistical differences between departments with high and low grievance rates typically appear to be meaningful even though the statistics can distort differences and always must be interpreted with great care. Consequently, only modest use is made in this report of statistical differences.

Reflecting the broad gamut of variables influencing day-to-day union-management relations, the investigation began with a loose hypothesis for differences in grievance rates, as contained in *The Impact of Collective Bargaining on Management:*

> When very low or very high grievance rates are found, they are usually associated with one or more of the following conditions: (1) the state of relations between the union and the employer, (2) the experience of the union and the employer in dealing with each other, (3) the personalities of key union and management representatives, (4) methods of plant operation, especially methods of wage payment, (5) changes in operating methods or conditions, (6) union policies, (7) union politics, (8) grievance adjustment procedures, and (9) management policies.[3]

These factors were classified, after abandoning more complex approaches, into a simple three-variable framework. Environment, union, and management were the variables taken as the primary elements; by using these three broad factors it was hoped that the more important subvariables impinging upon grievance activity and the resolution process would emerge, and that some judgments of their relative

---
[3] Ibid., p. 702.

importance would be possible. Although the three broad variables proved to be viable, and although it proved feasible to develop meaningful subvariables, the assessment of relative weight or importance to be attached to either the broad or narrower variables turned out to be essentially beyond the power of the research. The study had to be limited to a small number of departmental pairs to achieve some degree of depth in the investigation of each department. Such a study cannot reveal relative weights among highly interdependent variables except in an impressionistic manner.

In fact, each department studied was very much a unique situation with its own particular combination of variables that appeared to affect the rate of grievance activity. Despite this, some subvariables did appear consistently enough in either high- or low-grievance-rate departments to develop the high- and low-rate syndromes that are presented in Chapter 5. It has also been possible to make some judgments about the independence of some subvariables, such as union leadership, and the conditioning role that other subvariables, such as the community environment, play in the resolution process.

The research design was thus a simple one—the selection and study of pairs of high and low grievance departments in an attempt to highlight the variables contributing to the quantitative and qualitative differences in grievance activity. Through discussion with management in the five companies, departments were selected in six plants, ranging in size from 2,000 to more than 18,000 employees. In one company an essentially identical department was studied in each of two different plants. In the four other companies both the high and low grievance departments were from the same plant, thus holding some management, union, and environmental variables constant. Five of the six plants were organized by the United Steelworkers of America. In one plant the employees were organized by an independent local union. Although the selection of departments for study was done in consultation with management, there was no attempt, except in one company, to hold any particular variable constant. Exhibit 1 summarizes the companies and departments studied. Exhibit 2 gives the associated grievance rates. The names and other identifying features of all the companies have been disguised.

In Company 1, corporate management believed that the

Exhibit 1. SUMMARY OF COMPANIES AND DEPARTMENTS STUDIED

Company	Department	Operations performed
Company 1	Coke ovens (low)	Production of coke and chemical by-products
	Bricklayer (high)	Maintenance
Company 2	Steel mills (low)	Basic steel manufacture: open hearth, blooming mill, rod mill
	Wire mill (high)	Manufacture of wire and wire products
Company 3	Plant B (low)	Structural mill
	Plant A (high)	Structural mill
Company 4	Silicon (low)	Fabrication of silicon-steel sheets and strips
	Maintenance (high)	Plantwide mechanical and electrical maintenance
Company 5	Utilities (low)	Generation and distribution of electricity, steam, and water.
	Coating (high)	Coating and finishing cold-rolled strip

bricklayer department at the selected plant would be an interesting high-grievance department to study. Several low-rate departments were available to contrast with the bricklayer department and from them the coke ovens department was selected because it had a similarly high proportion of Black employees.

In contrast, at Company 5, though the high-rate department, the coating department, had a high proportion of minority employees (Black, Puerto Rican, and Spanish-American), the low-rate department did not. The proportion of minority employees was ignored in part because management did not believe that race or ethnicity had a significant effect in the high-rate department in this case. The low-rate utilities department was selected for study solely on the basis of its consistently low grievance rate.

The high-rate department at Company 3, a structural mill at Plant A, was selected by corporate management be-

Exhibit 2. GRIEVANCE RATES IN HIGH- AND LOW-GRIEVANCE-RATE DEPARTMENTS STUDIED[a]

Company	1964 High	1964 Low	1965 High	1965 Low	1966 High	1966 Low	1967 High	1967 Low
Company 1[b]	—	—	24.0	3.0	20.5	3.8	—	—
Company 2[c]	9.8	3.8	11.7	3.6	6.0	1.8	11.7	5.6
Company 3	2.4	8.1	5.1	4.5	8.0	9.5	18.0	6.9
Company 4	0.6	0.0	0.9	0.4	1.0	0.7	3.3	2.1
Company 5	20.0	3.6	11.5	3.7	31.2	2.6	24.6	1.6

[a] Grievance rates are expressed as grievances per 100 employees per year.
[b] Figures for 1966 were incomplete at the time of the study. The rates shown are minimum figures.
[c] A strike in 1966 affected figures for that year. The 1967 figures are fiscal year figures: July 1, 1966 to June 30, 1967.

cause they believed that the local union leadership warranted study. Since they also believed that it would be valuable to hold the workplace technology constant, a similar manufacturing operation in Plant B was chosen to contrast with the high-rate department. As a result, at Company 3, the factors of plant management and the local environment were different in the two departments studied, while in the other companies these factors were constant. When grievance rates were subsequently developed, a consistent statistical difference between the departments was not revealed (see Exhibit 2), although the quality of the union-management relationships clearly differed.

At Company 2 the focus of the study was the entire plant, with what at this company was called the steel mills (open hearth, blooming mill, rod mills) having a low grievance rate and the wire mill (fabrication of wire and wire products) having a high rate. Although the plant was not large, no attempt was made to study all operations in these two divisions, since a sample of three departments from each area was sufficient to reveal the differences.

At Company 4 the maintenance department was contrasted with the silicon production department. In these latter two companies, Numbers 2 and 4, the departments were selected on the basis of differing grievance rates, and no conscious attempt was made to hold any particular factors constant.

Written grievances per 100 employees per year was the measure of grievance activity for this study. The range in the rates, as shown in Exhibit 2, between the departments studied within a plant and among plants was large, and each department had large elements of uniqueness. At Company 4 the high grievance departments had a rate of 3.3 grievances per 100 employees per year in 1967, the highest rate that the department had ever had. The low grievance department had a rate of 2.1 per 100 employees per year. The rate in the high-rate department at Company 4 was lower than that in the low-rate departments studied elsewhere. The discussion in subsequent chapters will help explain these differences on intercompany grievance rates.

Some part of the variation in grievance rates measure chance differences in the filing and recording of grievances. The propensity of a union representative in a given location to file multiple grievances for a single event (for example, one for each employee involved), compared to the filing of a single grievance in another location, distorts the data in an undetermined manner. Individual grievance committeemen in a single location were not necessarily consistent on this score, filing one grievance in some instances and several in another. In other cases, grievances settled in an essentially oral and informal manner could, because of different procedures, end up recorded as a written grievance in one location and not in another. Differences in procedure also made the determination of the exact level of settlement difficult. Classification of grievances by subject also has its inherent difficulties. The statistics were developed on the scene at each of the plants, in some cases totally from raw data—that is, actual written grievances—and in others from company summaries supplemented by actual grievances. The statistics have inherent deficiencies, and the rates should be viewed as an approximate measurement. On the other hand, observation in the departments validated the reality of the major differences revealed by the statistics.

Differences in recorded grievance rates can be better understood and partially explained by contrasting hypothetical *challenge rates* and *problem rates* with the recorded *grievance rates*. Differences in departmental problem rates most likely reflect differences in the number of problems *created by environmental variables*. This is, of course, a special definition of what is meant by a "problem." The union leader, by way of contrast, would think of his "problems" more nearly as the product of the action or inaction of management rather than as the creations of environment. The management in turn would think of its "problems" as attributable in large measure to the character of union leadership. However, it is quite helpful analytically to define and associate differences in hypothetical problem rates only with differences in departmental environments.

As the analysis of the environmental variables in Chapter 2 will indicate, all of the high-rate departments—with the exception of Company 3, in which the environment was held constant in major respects but not in all factors—had higher hypothetical problem rates as herein defined than did the low-rate departments. In other words, the grievance rates were higher in the high-rate departments in part because the environments, particularly those in the operating arena, regularly and routinely gave rise to more problems than in the low-rate departments. Additionally, problem rates varied among both the high- and the low-rate departments. The low-rate departments in Company 1 (coke ovens) and in Company 5 (utilities) were markedly stable and in other respects free from environmentally created problems. The high-rate department in Company 3, selected for this reason by management, had not consistently over the years had a higher grievance rate than the low-rate department; this was in part because the differential impact of technological change had raised the rate in the low-rate department in some years to a level somewhat above what it might otherwise have been. Enough has been said, however, to indicate the analytical usefulness of hypothetical problem rates associated with environmental differences as a partial explanation of differences in grievance rates. The character of the environmental subvariables will be noted later in this chapter and subsequently explored in Chapter 2.

Another important element in explaining differences in written grievance rates can be visualized by contrasting hypothetical *challenge rates* with the actual *grievance rates*. Union representatives or employees, or both, may use an oral complaint or oral grievance to question or challenge the management's action or inaction. In fact the first step of the quite standardized grievance procedure in steel has an informal oral stage. If the grievance is not resolved in this oral discussion it is then put in writing and formally answered at the first step, essentially as a prelude for appeal to the second step. There may in fact be what amounts to an oral appeal procedure to the third step and resolution may be achieved through oral discussion, with no resort to a formal written grievance.

Most of the companies studied had a five-step grievance procedure. Step 1 involved the foreman and the employee or his grievance committeeman, or both, and was initiated orally but was converted to written form prior to formal appeal. Step 2 involved these individuals and the department superintendent. Step 3 was the top plant step and usually involved the chairman of the local union's grievance committee, committee members, and the superintendent of labor relations or his representative. The local union's international staff representative handled grievances at Step 4, meeting with corporate labor relations representatives. Step 5 was arbitration, with the individuals from

Step 4 representing the parties. At Company 2, a single-plant operation with an independent union, the standard fourth step was unnecessary.

Consistent with or somewhat independent of the five-step procedure there are various forms of oral and more or less informal, grievance activity. In the first place, employees may bring complaints or grievances to the foreman without involving union representatives. These complaints may be resolved or dropped and never enter into any recorded form. Although a considerable amount of this type of activity may exist, there was almost no evidence of it in this study. In the low-rate coke ovens department in Company 1 there was a record of this type of activity, and interviewing indicated that it was an important part of the problem-resolution process. However, the typical and essentially universal practice was for the union committeeman to be involved in anything that could reasonably be regarded as grievance activity. In the second place, some union representatives had a strong preference for oral resolution and reduced grievances to written form only as a last resort. In the third place, an oral appeal through company or company-union channels to the third step might occur. This informal appeal recognized that the company labor relations staff was the primary locus of authority in the disposition of grievances. Finally, at least in one company, there was strong management preference for oral resolution in association with active labor relations staff participation at the first step of the procedure. Thus various forms and degrees of oral and informal grievance resolution existed.

The hypothetical challenge rate—the sum of the written grievance rate and the oral grievance rate—can significantly exceed the actual recorded grievance rate. In addition, not only is the typical challenge rate higher in the high-grievance-rate departments than in the low-rate ones, but there tends also to be a higher degree of formal grievance activity with very little informal and oral resolution. A difference in the formality, quality, and typically, the level of the resolution process appears to distinguish high-rate from low-rate departments. The differences go well beyond the measured statistical differences in formal written grievance rates.

To distinguish between departmental grievance rates and hypothetical challenge rates and problem rates in grievance activity and the resolution process is helpful. There are substantial differences among departments in environmentally determined problem rates. Additionally, the challenge and grievance rates are most clearly determined not merely by the number of problems faced by the parties but, very importantly, by the character of the union-management relationship and by union and management leadership, organization, and policy. Some union individuals challenge management with very high frequency even though the problem rate is low and the problems and issues rather routine. Some managements and management individuals resolve problems with infrequent and informal union challenge even though the problem rate appears to be rather high.

Although it is useful to conceive of environmental forces as responsible for the number and character of problems faced by the parties, and union and management organizations and individuals as important additional determinants of the challenge and grievance rates and the resolution process, this introductory point of view oversimplifies the case. As the study progressed it became clear that the variables were highly interdependent and complex. For example, the environment unquestionably and importantly conditioned the resolution process as well as the challenge and grievance rates, as the following brief discussion of the analytical framework indicates.

THE ANALYTICAL FRAMEWORK

Environmental forces and conditions are difficult to classify adequately since they are so numerous and diverse. Also they are especially difficult to assess when they behave not only as active forces but as passive conditioning influences. However, the more important broad environmental influences may be classified under the following headings: (1) socioeconomic conditions, (2) technological change, and (3) task organization and the work environment. Though all three categories are important, the ramifications of the third category seemed more impressive. Variations in routine operating processes in the departments created substantial differences in the number and character of the problems and conditioned the resolution process.

As socioeconomic variables, general business conditions were not a source of significant problems at the time of this study. Local business conditions did vary, for some local labor markets were considerably tighter than others. Some problems were created by unfilled vacancies and by variations in the qualifications of new hires, but in general, business conditions were a favorable influence.

The civil rights issue was the salient social variable. Many departments studied reflected to some extent this widespread social problem, but the degree to which particular departments were affected was not subject to simple explanation.

Size and location of community, plant, and department appeared to influence the climate of relationships in the department. The size of management and union groups, the presence or absence of contacts outside the plant, and the length of internal lines of communication appeared to be significant. It was not possible in this study to probe social variables in great depth.

Influences of technological change are widely recognized. Such change in addition to creating layoffs, transfers, promotions, and demotions alters job content—with consequent implications for wages and other factors. What was especially interesting was that a history of job insecurity in a department and the uncertainty of future impacts tended to change the climate of the relationship quite apart from the direct creation of particular disputes.

Of particular importance were differences in task organization and the work environment. The degree of attention required, the repetitious or variable character of the work, the individual or group nature of the task, the location and mobility of the work, the skills required, and the favorable or unfavorable working conditions all served in various combinations to create widely differing work environments. In basic steel these differences had a special effect on the operation of incentive plans. The number and character of problems arising in different departments under normal operating conditions was an important variable. It is of equal significance that some working environments were highly conducive to informal resolution of problems while others were not.

It is hoped that what has been said indicates the broad outline of environmental influences, their relation to a hypothetical problem rate, and their potential impact on the formality of the resolution process. Actual grievance rates do reflect these environmental differences, but only partially. One of the most stable environments encountered, which presently has a low grievance rate, at one time had the highest grievance rate in the plant. In brief, union and management variables can considerably alter environmental expectations.

The union and management variables are each analyzed in terms of the following criteria: (1) leadership, (2) organization, and (3) policy. So far as the union is concerned, these three subvariables, although related, are viewed as quite independent. Differences in union leadership behavior were the most conspicuous variable in the study. The range of variation in both union and management organization and policy was limited because the study concerned only one international union (except in one company) and only one industry, with a centralized contract negotiation process, the result being almost identical basic agreements in each company and considerable similarity among companies in policy and practice.

A variety of union leadership styles and underlying motivations were observed. The chapter on union variables attempts to classify leadership styles in relation to workplace problem-solving activities. In the union analysis, the leadership variable stands out as highly important.

Some effects from internal union political activity are noted in the union chapter and discussed in association with union leadership. In a study contrasting departments, union politics was not revealed as an important variable apart from differences among union leaders. In a grievance study contrasting plants or companies, a variable to indicate union political stability or instability might well be significant.

Union organization was similar and not particularly complex in plants organized by the Steelworkers Union. Although power was shared among a number of individuals the locus of power in each department tended to be concentrated in a single individual—most commonly the committeeman or grievance committee chairman, who might also be a local union officer. Lower level grievance activity tended to be dominated by that single individual, accentuating the importance of the particular individual's style of leadership. Union stewards had no meaningful grievance role. The plant not organized by the Steelworkers had a multiple committee structure to process grievances, creating a most complex resolution process with considerable diffusion of power.

Union policies revealed by the study were quite specific and played only a modest role in the grievance process. Where such policies were used, cases involving certain issues were consistently put forward, and nearly always carried to arbitration.

The management variables of organization, leadership, and policy are more interdependent than in the union situation and no single subvariable stands out conspicuously. For example, within management, decision-making authority is much less apt to be concentrated in one individual than in the union, and individuals are more subject to organizational and policy constraints. Generally speaking, the organizational and policy variables appeared relatively more important and leadership less important than in the union analysis.

Line management organization does not reflect the relative difficulty of the managerial task that exists in different departments. The most important aspect of managerial organizations is the line-staff relationship. This relationship varied significantly in different companies even though, at the higher steps of the grievance machinery, decision-making authority in all companies rested with the staff. The differences in line-staff relationships appeared to be very important. Easy generalization, however, is impossible. This question is given considerable attention in Chapter 4.

Three selected management policy considerations were singled out for analysis: consultation, progressive discipline, and incentive management. In general, lack of policy is conducive to grievance challenge, but it is also clear that strong and frequent challenge makes consistent policy application extremely difficult.

Several types of managerial leadership were observed. In general, foremen were seen to have virtually no influence in the resolution process, but those managers who played an important role in the process are classified in Chapter 4 in terms of the nature of their response to the union grievance challenge.

Research Methodology

The research demonstrated the usefulness of the three-variable framework in revealing the full dimensions of departmental grievance activity. The study itself began with the selection of high- and low-grievance-rate departments. The next step at each location was the careful perusal of written grievances. The statistics, although imperfect, provided a guide to the amount of grievance activity and a converse indication of informal problem solving as well. Classification of grievances also indicated the subjects and problem areas of greatest concern.

The careful examination of grievances brought the first familiarity with the names of both union and management personnel, as well as increased familiarity with specific problems in each department. The nature of the union-management relationship was also apparent in the written grievances. Charges of discrimination and persistent violation were immediate clues to specific areas of potential conflict within a department. The examination of arbitration cases—the arbitrator's decision and the materials developed by management and briefs submitted by both parties—developed familiarity with major problems faced by the parties in the past, which in some cases involved issues which were still troublesome.

The investigation of grievance records was supplemented, on occasion, by records of informal settlement through discussion. Such records were used at Companies 1, 2, and 5, but not all informal discussions were placed on file. Even so, the records provided a rough indication of the amount of informal problem solving. The low-grievance-rate departments in each of the companies using these records evidenced more informal discussion and problem solving than did the high-rate departments.

The examination of the pertinent available records preceded the central part of the investigation—the interviewing of management personnel, union officials, and employees. This interviewing was done primarily at the department level and focused on the foreman and general foreman, although it included all levels of line management. The interviewing also included staff managers in both industrial engineering and industrial relations. Interviewing in the union organization paralleled that done in management. The department grievance committeeman, the chairman of the local grievance committee, and on occasion, the international staff representative were all interviewed.

The size of the departments investigated ran from 280

to 1,416 employees, and the size of the management group varied accordingly. Generally speaking, it was possible to interview all members of the departmental management group down to the foreman level. At the foreman level simply too many individuals, working on several shifts, made it impossible to cover this group adequately, so a sample of foremen were interviewed.

These foremen, and usually the general foremen, too, were interviewed for brief periods on a daily basis. In addition to specific questions on issues that the written records indicated would be profitable lines of inquiry, the foremen and general foremen were asked to recall requests or complaints by employees and/or their union representatives, or both, which had occurred either that day, or since the time they had last seen the interviewer, or simply, recently. Questions were also asked about the role of the union representative, the management chain of command for the resolution of formal and informal problems, disciplinary procedures, particular problem areas, and finally, the supervisor's perception of his role as manager and his reaction to the problems he faced.

The interviews with other line managers at higher levels of the organization tended to be fewer in number but somewhat longer. The same basic questions were asked. The interviews with union officials and with management staff people other than labor relations were similar to those within higher levels of department management. Because the researcher's introduction to the plant and the departments under investigation was through the labor relations department in each of the locations studied, interviews and discussions with labor relations personnel were generally on a daily basis over the entire length of the stay in each place, which was from four to six weeks.

The final tool of research, in addition to the study of records and interviews, was observation. In the course of workplace interviews it was possible to observe the nature of the work being performed in the department as well as the requirements of the different jobs or tasks which the work was divided into. Many of the interviews were done "on the hoof," in the course of following the foremen or general foremen as they went about their tasks. And at the same time the nature of the supervisory task and some of the activities performed by supervisors were also observed, as were interactions between supervisors and their superiors, subordinates, peers, and various staff groups.

Grievance meetings were observed in every location. In general, meetings involving line management and the union were attended as well as those with labor relations staff and union representatives. The former were usually second-step meetings, involving the superintendent and the department grievance committeeman, although on occasion a first-step meeting was observed. The latter meetings were third-step meetings, usually including the plant labor relations manager or his representative and the chairman of the local's grievance committee.

In all of the interviews and observed interactions the line managers, staff, union officials, and employees were cooperative, candid, and open. In those rare instances in which individuals were reluctant to talk at all, a second or third visit almost always brought an increased willingness to discuss problems. All information was not taken at face value. Interviewing both union and management, as well as line and staff, and different levels of line management provided an opportunity to check each story, to hear the other side. This was done regularly in the course of the interviews, and each interview provided new questions for the next.

This then is the research activity that was undertaken to develop information on the causes of differences in grievance rates, on the factors that influence the level of settlement, and the roles the parties play in the resolution procedure. This research effort, which meant virtually "living" in a department, developed a great quantity of information. The findings at the various companies are greatly compressed in the following three chapters.

Each of the next three chapters concentrates on one of the variables. Chapter 2 is devoted to environmental variables, Chapter 3 to the union variables, and Chapter 4 to management variables. Chapter 5, the concluding chapter, after a synoptic review of the study, analyzes the relationship and interdependencies among the variables and presents a view of high- and low-grievance-rate syndromes. Finally, some of the implications of the study for management are discussed.

2 Environmental Influences on Problems and Their Resolution

IN OUR THREE-VARIABLE MODEL, environmental factors can be described as the context within which the other two factors, union and management, operate. The union would perceive itself as operating in the context of a given environment and a given management. Management would see its "givens" as the union and the environment. The concept of environmental factors is a difficult one because of the complexity of the world in which union and management relate to each other. Before the study began, the variety and scope of the influence of environmental factors on the generation and resolution of problems, did not seem especially important. Interestingly, the labor and management participants in the resolution process also tended not to be particularly conscious of the environment. The parties tended to take the environmental dimensions of their situations for granted, to treat them as given parameters within which they had to operate.

The methodology used for this study forced a reconsideration of initial views on environmental influences. By simply using differences in the grievance rate as the basis for selecting departments for study, a variety of environmental conditions were simultaneously offered for investigation. Clearly some environments routinely gave rise to many more problems than others. Also, in moving from company to company and plant to plant, differences in the larger social and economic contexts were observed. Although these broader influences—with the exception of the civil rights issue—did not seem to have so dramatic an impact as the factors closer to the workplace, they did have some power. These broader influences might be more important in other industries or might have been revealed more clearly by a study allowing trend analysis. Nevertheless, one of the strengths of this study is that it illustrates the competence of *both* environmental and behavioral factors to explain departmental differences in the administration of collective agreements.

The environmental factors that were outlined in Chapter 1 will be discussed here in greater detail under three broad categories: (1) socioeconomic conditions, (2) technological change, and (3) task organization and work environment. The economic part of the socioeconomic environment includes general business conditions, local economic conditions, and shifts in markets and product mix. The social factors include general social conditions, particularly civil rights activities and local social variables relating to the size of the community, the size of the plant, and the size of the department. Technological change, though conditioned by economic forces, can appropriately be discussed separately. Differences in task organization and the work environment are a reflection of variation in the technology of production processes and also warrant separate discussion.

Socioeconomic Conditions

General economic conditions had a direct impact on the workplaces being studied. The most immediate effect was on job security. Cyclical downturns or changes in the demand for certain products can be the cause of layoffs. These variables operate through management and union organization and policies. A cyclical downturn means layoffs only to the extent management acts to furlough employees, and such an act creates a problem for resolution only if the union challenges the action, presumably on the basis that it was not done according to pre-established rules.

The research in steel was conducted between August 1, 1967, and October 1, 1968. This period was one of continuing prosperity and low unemployment in the economy as a whole. The three-year basic steel agreement was due to expire on August 31, 1968, and production was heavy before that date in anticipation of a strike which did not materialize. The combination of generally favorable economic conditions and prestrike conditions typically meant that the plants studied were operating at high levels of output, if not at full capacity. No observable differences could be directly related to the effect of gross economic conditions. Presumably if economic conditions had been deteriorating, the grievance rate would have been affected by an increased number of seniority-related problems. The study at Company 5 was conducted after the successful conclusion of negotiations. Although production had declined at the plant, the departments studied were not deeply affected. Still, there were some problems with layoffs and with new job combinations that postdated the new agreement.

The high level of economic activity may have meant that supervisory personnel were more interested in getting out production than they might have been with less economic compulsion, and consequently they may have had less time available for informal problem resolution. Also, they may have been more susceptible to the use of pressure tactics. But this is all conjecture. The only established fact is that at the time of this study economic conditions were good and production was at high levels. Under different circumstances, the observations and conclusions reached herein might have had a somewhat different emphasis.

Although the overall economic environment was relatively constant throughout the study, there were differences in the local economic environments in which the plants operated. For example, Companies 1 and 2 were operating in extremely tight local labor markets, while Company 4 was not. The most visible difference was seen in Company 3, between Plant A, located in a large metropolitan area, and Plant B, which was in a much smaller community. At the time of the study the labor market in which Plant A was located was extremely tight. Many times during the study at Plant A there were no laborers available within the structural mills—the department under study—since all available men had been pulled up into specific machine crews. It also meant that the performance record of newly hired employees in the Plant A structural mill was much lower than in the Plant B mill. Plant B was located in a predominantly agricultural area with a surplus of potential employees. With few exceptions, Plant B had not hired a new employee without a high school education in the ten years prior to this study.

Again the implications of these differences are difficult to pinpoint. Although the Plant B structural mill did not have a consistently lower grievance rate than its counterpart in Plant A, the quality of the procedure and of the union-management relationship was superior in Plant B. This difference is primarily attributable to variables in the local unions, as will be seen later, but the character of union leadership and the union's policies and practices may well have been, in part at least, a reflection of its constituency.

Two other types of economic pressures should at least be mentioned here, although much direct evidence for them is not available from this study. The first concerns the westward shift in domestic steel markets. In one location studied, Company 1, this shift would probably mean a reduction in raw steel output at the plant, as a new plant farther west began producing its own steel for fabrication. Job security was reduced accordingly in one of the departments studied and was reflected in the grievance rate. Further details are provided later in the chapter in a discussion of the impact of technological change, which was also important in that department.

Foreign competition was one of the compelling reasons why Company 2 had embarked on a major series of technological improvements that had not been implemented at the time the study was completed. The company was under severe pressure from foreign-made steel products. However, because the company had only one integrated plant and a single product line, the consequences of this pressure did not have any differential effect on job security and grievance rates in the departments studied there. Foreign competition had uniformly reduced job opportunities and had a significant impact on profits. In a plant with a multiple product line, foreign competition might well have an impact on only some products and this in turn would have an effect on the relative grievance rates. Economic conditions, both national and local, can thus be seen to have the potential for affecting the climate of the union-management relationship and for influencing the problem rate and the number of challenges to management decisions as well as the way in which those challenges are resolved.

Even more important than local economic conditions were the size and nature of the community in which the plant was situated. The size and agricultural orientation of the area in which Plant B of Company 3 was located influenced not only the local labor market, but the relationships within the work force as well. In this kind of smaller community, managers and employees tended to have more associations outside the plant environment, and as a result tended to know each other better. Informal settlements were thus made easier.

The plant of Company 4 was located in a very small town, and the local environment appeared to have the same beneficial effect on the relationship. Company 2 was located in a larger, more industrialized area than Plant B of Company 3 or Company 4, and the potential for frequent outside interaction did not exist. However, more opportunity seemed to be found at Company 2 for personal outside contacts than in the locations of Plant A of Company 3, Company 1, and Company 5. In these large urban centers, the work force tended to be scattered over a wide area, and management and employees tended to live in separate communities and had separate interests outside the plant.

The lowest overall *plant* grievance rate uncovered in the course of this study was found in Company 4. Not only was this plant located in a small community, but it was also relatively small. Although the size of the plant was not a direct determinant of the relative grievance rate, it did influence the length of internal lines of communication and the number of the management group as well as their accessibility. The low grievance rate at this Company 4 plant can more properly be attributed to the line-staff union relationships within the plant (which will be discussed in detail later), but those relationships were, among other variables, a function of plant size.

The size of community and plant do not of course affect the relative grievance rates within plants, but it appears that differences in the grievance rate between plants can be influenced by these factors. The smaller the plant and the smaller the community, the lower the grievance rate is likely to be.

Both social and economic factors affected the size of the plant and the size and nature of the community. One particular social factor—civil rights—was decidedly important and requires special mention. The impact of the civil rights movement could be seen in several of the work places studied, and its impact at work was a reflection of its importance as a social issue and problem in the wider community.

The impact of the civil rights movement was obvious in the bricklayer department at Company 1. The overriding concern of the chief shop steward (grievance committeeman) of the bricklayer-laborer group was civil rights, or discrimination. Of the many grievances he filed, most involved or included charges of alleged discrimination by supervisors against the primarily Black laborer group. The bricklayer laborers were a relatively low-status, low-pay group working closely with the higher status, higher paid White bricklayers. For Black helpers the path to becoming bricklayers was blocked at this time because of their inadequate educational qualifications, and because apprenticeship opportunities were limited by work force reduction due to technological change. In this manner, work place conditions reflected conditions in the wider community, with the Black employees aspiring to more than was currently available to them. Under militant leadership the grievance procedure became one outlet for their frustrations, even though the real problems—the latent content of the grievances as opposed to the manifest content —could not possibly be resolved through that procedure.[1]

The proportion of Blacks was similarly high in the low grievance department studied at Company 1, the coke ovens department. Here the highest status and the highest paying jobs in the department were and had been open to Blacks. This was primarily because a large number of Blacks had traditionally worked in the department and because there were no educational bars to individuals in occupying the top jobs, which was not the case in the bricklayer department. Militant leadership also was not a consideration. Although the situation in the bricklayer department had existed for some time, intensification of the problems management faced in that department at the time of the study clearly reflected the civil rights pressures

---

[1] F. J. Roethlisberger and William J. Dickson, *Management and the Worker* (Cambridge, Mass.: Harvard University Press, 1939). See the classic analysis in Part III, pp. 255–376.

in the wider community. The contrast between departments was marked.

Company 1 was the only location in which the civil rights issue was reflected in interdepartment grievance-rate differences. In another of the plants studied, but some time prior to this study, the company merged the separate seniority systems that had been maintained for Blacks and Whites. This led to some confusion and to some grievance activity.

Human rights issues and problems are of course not unique to the steel industry. In the automobile industry, civil rights problems have caused administrative difficulties for both union and management. In Canada, the issue of French Canadian civil rights had meant similar problems for companies operating in Quebec, and these problems have been reflected in grievances. Similarly, difficulties associated with women's rights have found their way into the negotiation and administration of collective agreements.

Thus social problems in the wider community will be reflected in the industry and will affect the grievance rate between plants and between departments to the extent that these wider social problems exist, in microcosm, at the plant level. At Company 1, the frustrated aspirations of Blacks in one department with militant leadership, led to a grievance rate significantly higher than in the other department studied. In another plant, correcting past discrimination by merging seniority lists led to an overall (but temporary) increase in the grievance rate. The emergence of social issues is in some instances an important conditioning variable of the grievance rate.

In summary, socioeconomic conditions tend to influence the climate of the labor-management relationship and particular aspects of issues. The specific application of these general variables to the work place is through union-management organizations and policies. The civil rights problems of the larger environment were reflected in varying degree in a given department by the union leadership. The social aspects of the size of the community and the plant affected internal lines of communication and the relationship between the parties. Although the impacts of these variables have been somewhat difficult to pin down, they do exist and they do influence the climate and content of the labor-management relationship.

To a great extent socioeconomic variables may tend to influence interplant rather than interdepartment labor-management relationships and grievance rates. Civil rights and related issues may have different effects, and economic pressure, insofar as it relates to specific product lines, may be reflected in differing grievance rates. In general, though, the socioeconomic variables discussed above were most apparent in interplant rather than interdepartmental comparisons. These factors influence the grievance procedure, but the remaining environmental factors have a more direct impact on interdepartmental grievance-rate differences.

TECHNOLOGICAL CHANGE

Technological change, fostered by economic pressures, had an immediate impact on the place of work. Although the effects of cyclical economic forces were not particularly evident during the course of this study, the present and probable future effects of technological change were. Technological change affected job security and job content, where, in addition to job elimination, it acted to create and modify jobs, leaving an opportunity for subsequent disagreement over classification, pay, and manning.

One of the most significant technological changes in the United States steel industry in recent years has been the replacement of open hearth steelmaking by the basic oxygen furnace techniques. The study included one open hearth operation in which the introduction of electric furnaces was just being discussed. No study was made of the new basic oxygen furnace operations in the other places, so the direct impact of this new technology is not part of the evidence in this investigation. (The classification of basic oxygen furnace jobs was the source of an industry-wide dispute between the companies and the Steelworkers for several years, a dispute which was resolved during the 1968 contract negotiations.) The impact of the installation of basic oxygen furnaces was felt, however, by one of the craft groups studied.

In the high-grievance-rate department of Company 1 a large part of bricklayer work centered around the relining and rebuilding of open hearth furnaces. The installation of basic oxygen furnaces eliminated the use of fourteen open hearth furnaces. Before the installation of the BOFs, the bricklayer department employed 200 bricklayers and 600 laborers who worked six- and seven-day weeks with some frequency. Bricklayer department employment at the time of the study was 123 bricklayers and 373 laborers. The BOFs were installed in 1964, but the actual decline in bricklayer department employment occurred in 1962, due to a decrease in the overall level of business. The installation of the BOFs simply meant that employment never picked up as business improved. The decline in employment was directly associated with the BOFs by department management and employees.

The job of relining BOF vessels was done by bricklayer department crews. However, although the amount of steel produced in BOFs was greater than in open hearths, less frequent repair work was needed, and fewer men were required for the BOF relining process. Also, some of the teardown and materials-handling work done by bricklayer department employees on the open hearths was performed by BOF department personnel and other crafts in the new operation.

BOF vessels needed relining less frequently because of the superior refractory material (brick) used as lining. The improvement in refractory material and the use of nonrefractory materials which were easier to apply had been going on more or less continuously, and its impact was difficult to assess. However, this improvement was noticed by department employees. One effect of the use of new materials was to reduce the skill levels of bricklayers. This meant a narrowing of the perceived skill differential between the White bricklayers and their Black helpers, compounding the difficulties that stemmed from the helpers' unmet job aspirations.

As a further threat to job security in the bricklayer department, approximately one-fourth of the basic steel output of the plant was being shipped to another of the corporation's plants which had no basic capacity of its own. That basic capacity was under construction at the time of the study, however, and the bricklayer department employees knew that in the absence of improved business conditions, production of steel at the plant was likely to be reduced when the new capacity at the other plant went on line.

In practical terms the workers' concern in this department with job security could be seen in the kind of grievances. Almost 40 percent of the written grievances involved seniority, subcontracting, and work jurisdiction issues. These jurisdictional problems were twofold. One type of problem involved disputes among crafts, for example, bricklayer, millwright, rigger. The other type involved complaints by the laborers that they were doing craft work, for example, bricklayer work, and should be paid for it. A brief wildcat strike during the study centered around a refusal by the bricklayer laborers to build a scaffold unless they were paid rigger's wages to do so.

The impact of technological change was not visible in any of the other high-grievance-rate departments studied. Some important equipment changes would take place at the high-rate Plant A of Company 3 after this study was completed there. However, the impact of these changes was on the agenda for negotiations later in the year and had not yet been apparent in on-the-job behavior, except for references to the new equipment in daily conversations.

The other manifestations of technological change appeared in low-grievance-rate departments. Before reviewing the specifics of these situations two factors can be noted in connection with the appearance of problems associated with technological change in these low-rate departments. First, one of these departments—Plant B of Company 3—did not have a consistently low grievance rate, particularly in comparison with Plant A, and technological change partially explains why the grievance rate was not consistently low. Second, in some instances grievance rates can remain low in the face of technological change and other problems because workers choose not to resolve these problems through the formal grievance procedure. They resort to other methods, principally pressure tactics.

For example, at Company 4, the low-rate silicon department was the center of technological change in an entire operation, the manufacture of silicon steel. But this operation was in the process of being transferred to a physically distinct location. The new silicon department began operating as pieces of equipment went on line. The new equipment performed steel-finishing functions in an improved manner and generally required fewer employees than the old operations. However, job displacements had been a relatively minor problem. Also, this was a new department, and the employees not only had *not* been displaced, but in some instances, at least, enjoyed *greater* relative seniority than formerly through bidding on jobs on new equipment for which they were, at the time of the study, the only qualified operators.

The struggle in the silicon department was over incentive rates. The employees generally did not use the formal grievance procedure in this dispute but rather emphasized pressure tactics in the form of slowdowns. Problems with incentives typically related to technological change and will be discussed in detail in Chapter 4 under management incentive policies.

Although incentives were the main issue in the silicon department, other issues relating to the newness of the operation did exist and were handled through the grievance procedure. In fact, almost all actual written grievances were related to the change in the operation, specifically manning and job description and classification. Several of these grievances went to arbitration.

At Company 2, technological change was imminent in the low-grievance-rate steel mills part of the plant, but no effects were evident at the time of this study. The construction of a continuous casting operation was almost complete at that time. The installation was being built on a turnkey basis, and Company 2 personnel had generally been barred from the new facility. Union and management were beginning to discuss the manning of the new facility at the time of the study, but no agreement had been reached.

Rumors circulated in the plant at the time of this study that the company was going to begin operating the continuous casting facility using supervisory personnel only. This was strictly a rumor, since the company had no definite plans to do so. However, some union officials reportedly told management that any such attempt would result in a walkout by personnel at all the steel mills. This threat indicated that decisions over the manning of the new operation would be difficult, but also reflected union uneasiness over the ambiguity of the technological change and its future impact. The threat was probably not idle and indicated that employees might well resort to pressure tactics rather than the grievance procedure to solve any dispute which arose over the introduction of more modern technology.

Further impending technological change in the steel mills department at Company 2 included the installation of electric furnaces, which might possibly displace personnel in the open hearth operations; but at the time of this study construction had not yet begun. Also, at Company 2 the seniority system at the steel mills operated to provide more job security in the face of technological change than did the seniority system in the wire mill. The steel mills had essentially a plant seniority system, while the wire mills had a department seniority system.

The effects of industrial change could also be seen at Plant B of Company 3, another low-grievance-rate department. In Plant B the job security problems were an outgrowth of the installation of basic oxygen furnaces. The appearance of these problems among employees assigned to the structural mill was related to the seniority system. The plant had a single labor pool which was divided into three zones. These zones corresponded to three sets of operating units. Employees bumped out of a regular job had first chance at the labor pool jobs in the zone that was related to their regular job and department. If there subsequently remained unfilled pool jobs in a zone, then available employees on layoff from operations in the other zones filled them. The local agreement that established the three-zone pool had a strong "ability to do the job" provision; in addition, under this provision the company was not required to assign an employee to a job until thirty days after his layoff.

At the time the study was made, no employees were on layoff but the union was objecting strongly, in part through the grievance procedure, to the assignment of men to jobs within a pool and then from the pool to the bottom jobs on seniority lines. These difficulties helped make the grievance rate in this low-rate department higher at times than its high-rate paired-comparison department. The proportion of seniority grievances in Plant B was much greater than in Plant A. Because of the strong ability provisions, the company had been able to place men on particular jobs on the basis of experience rather than continuous service. Such assignments were frequently made because supervision preferred to place a man whom they knew had experience on a job rather than finding and training the man with the most continuous service. The union objected when continuous service was not followed, especially since many of the employees working in the labor pool were former open hearth production and maintenance employees displaced when the open hearth was shut down at the start-up of the BOFs at Plant B three years prior to the study. Under the operation of the seniority system, these men were unable to get permanent work on the bottom jobs of seniority ladders and continued to suffer a large cut in income because their work was at a lower job classification and without incentive earnings.

The continuing problem of the employees displaced by the use of the basic oxygen process was evident in the structural mill because of the seniority system and because of union organization. The impact of the installation of BOFs in Plant A did not filter down to the structural mill because the displaced employees had a separate seniority system and were represented by a different local union. In Plant B there was only one local union, and internal pressure forced the local to be concerned with the assignment of former open hearth employees in the structural mill department. The multiple-local organization in Plant A did not face the zone and job transfer problem.

Thus the union organization—whether a single local or a multiple-local plant—affected the grievance rate caused by problems of job security and job assignment associated with technological change. The presence of multiple-local unions was in this study a function of plant size, which in turn was related to the socioeconomic factors discussed earlier.

Worker dissatisfaction over the seniority system and its operation in Plant B was not limited to grievance activity. Workers believed that the negotiation of the seniority agreement represented a serious blunder by the local union leadership. That leadership was unseated following an election campaign in which the seniority system was a major issue.

The presence of technological change has been seen to affect the grievance rate through the creation of issues such as manning, job description and classification, work assignment, and relative status seniority. It has also been seen as an issue behind the use of pressure tactics. The presence of technological change can frequently be said to be an element in a high-grievance-rate situation, but it is by no means totally explanatory. A good bit of the data on technological change derives from low-grievance-rate departments. Certainly, without the problem associated with industrial change, the rate would have been lower in some low-grievance rate departments.

The effect of the *absence* of technological change is difficult to assess with certainty. Aside from a reduction in the number of problems subject to grievance activity, the stability resulting from the absence of technological change can be said to influence the climate in which the union and management operate. The impact of the absence of technological change and the presence of other factors which tended to enhance job security can best be seen in the coke ovens department at Company 1. This low-rate department contrasts particularly well with the high-rate bricklayer department at that location.

The employees in the coke ovens department at Company 1 had much greater job security than their counterparts in the bricklayer department. Coke ovens department employees had greater job security for reasons just the opposite of those inducing job insecurity in the bricklayer department. No new equipment had been installed in recent years, nor had there been any significant modification of existing equipment. No changes were planned either. Also, the requirements for coke at the plant were greater than the production capacity of the coke ovens department. At peak operation the coke ovens were capable of producing only 80 percent of the coke required for blast furnace operations. At the time of the study the BOFs alone were using the entire hot metal output at the plant, and the open hearth furnaces in operation were being charged with cold metal. Consequently, the prospect of a reduction in the output of basic steel would not in all probability have meant a reduction in coke production and coke ovens department employment. During periods of cyclical downturn, employment in the department had frequently not been affected, as unconsumed coke could be stored to reduce purchases when full production resumed. Production for inventory during slack periods was also attractive because the coke ovens were not easily stopped, cooled, and restarted, at least not without damage to the equipment.

Except for promotions and transfers, there were no potential problems with the seniority system in this department. Nor were there questions of manning or job classification. When combined with capacity deficiencies, a stable task environment (which will be discussed next), the absence of civil rights issues, the absence of technological change in this department meant an almost problem-free environment which contributed significantly to the low-grievance-rate syndrome.

Variables discussed to this point are well recognized in the literature. Economic and technological change have an impact on job security and individual wage rates. The frequency and severity of these changes create problems that give rise to grievance challenge. Also the significance of the social variables in conditioning and influencing challenge are frequently noted. Where economic, technological, and social changes have direct employee consequences, the association with grievance activity is obvious. Where economic, technological, and social factors are stable but operate as conditioning influences, their favorable effects are necessarily more speculative. But stability contributes to low grievance activity and the social climate in small plants and communities appeared to affect favorably the resolution process.

The variables to be discussed in the next section are very important. The number and character of problems arising in the normal and routine operation of a department vary enormously with differences in task requirements and surrounding conditions. It is these differences that will now be discussed.

Task Organization and Work Environment

In addition to problems created by changes in the technology of the work place, problems are created by normal operating requirements and conditions of the job and its location. Seven specific types of differences in task characteristics between high- and low-grievance-rate departments were seen particularly to affect the grievance rate, the resolution process, and the character of the grievance issues.

The first difference involved the degree of attention the operator was required to give to his work, from constant to intermediate to sporadic. The second related to the degree of control required by the operator over the quality and quantity of work. A third difference was whether the work was varied or repetitious. A fourth element was whether the work was done by individuals or groups of individuals (differences here being reflected in the amount of coordination involved between individuals and groups as well as the opportunities for personal interaction). A fifth difference related to the locus of the work, whether it was concentrated in one location or done in a variety of locations. A sixth difference was whether jobs were skilled, semiskilled or unskilled, and this difference was reflected in the characteristics and training of the people employed. The seventh and final difference was in working conditions, which ranged from relatively favorable to quite unfavorable.[2]

The differences in task organization and environment were not found in isolation in the departments studied. Departments with differing grievance rates typically diverged along several of the dimensions listed above.

Illustrative differences among the departments will be presented to develop an appreciation for the importance of these variables. Though reference will be made to the paired department, low-rate departments will first be described and then high-rate ones. Since at one company the same department was studied in two different plants, four low and high pairs are available in which to contrast task requirements and operating conditions. Special attention will be given to the effect of differences on incentive payment because of its importance in the steel industry and because of the direct and important connection between task requirements and the number and character of problems associated with incentive payment administration.

At Company 5 the low-rate department was the utili-

---

[2] John T. Dunlop, *Industrial Relations Systems* (New York: Henry Holt and Company, 1958). See chap. 2 on the technical context of the work place. The variables discussed by Dunlop include factors similar to or identical with those noted above. This research owes much to Dunlop's framework of analysis. Our debt to him goes well beyond the discussion of the technical context of the work place.

ties department (paired with the coating department, which coated cold-rolled strip with a variety of materials). The utilities department was an internal service department which produced electricity and steam, pumped water, and handled waste materials for the entire plant. Generally speaking, a production job in the department included the monitoring of automatic equipment and occasional adjustment of switches and valves. The work involved very little physical effort, had low attention requirements, and almost no quality and quantity pressure on the employees. The product was homogeneous, and the critical variable was simply its availability, not its quality. All of this was in marked contrast to the coating department, where the product was physically handled by production employees under strict quality requirements, particularly relating to surface characteristics.

Utilities department work was conducted in 40 different locations throughout the plant. Because of this dispersion and the character of the work, supervision was relatively light. In fact, supervision was present on only one shift. The work required quite a high caliber of employee. This was reflected in the classification of jobs in the utilities department, which were listed at a higher level than those in the coating department. Utilities department employees were older than those in the coating department, and the work force in the utilities department had a much smaller percentage of minority group members, primarily a reflection of the educational requirements. In brief, task requirements and conditions of work in the department were highly conducive to a low grievance rate.

Of all the departments studied, the utilities department at Company 5 presented the most favorable task environment. Very few problems were generated in this operating context and there was very little focus for employee or union-management conflict. Little physical effort was required. There were no quality demands. There was no incentive system of wage payment although wages were relatively high. Conditions also facilitated the informal resolution of such problems that did arise.

The coke ovens department at Company 1, a second low-rate department, was paired with a maintenance department. The central part of the coke ovens department was the actual coke-producing mechanism—the coke ovens. In these batteries of ovens, coke was produced from coal. Coal was introduced into the ovens from the top, and coke was pushed out from one side of the oven into a car on the other after a coking period of approximately seventeen hours. The sequence of operations was as follows: (1) remove doors, (2) push out coke, (3) carry coke to quench station (where it was sprayed with water to extinguish it), (4) dump coke, (5) replace doors, (6) refill oven with coal. This was the task of a crew, and this sequence was performed every ten minutes, twenty-four hours a day, seven days a week. It was performed in the same place—movement to a different battery of ovens by an employee was infrequent—and under the same supervision. Exceptions to the routine were infrequent. High stability in the production routine plus employment stability previously noted and almost constant incentive earnings, because of the technology, were clearly conducive to a low grievance rate.

The incentive payment in the coke ovens was hardly a true incentive in that it had a minimum influence on production. Though payment was based on output, output was almost totally regulated by coking time, and thus incentive earnings were very stable. The primary rationale for the coke ovens incentive plan was to reduce the impact on relative earnings of high incentive earnings in other units such as rolling mills. In modest degree the incentive encouraged full utilization of equipment with a minimum crew.

The economic and technological environment in the coke ovens brought security and stability. At the same time it did not create high pressure on employees for quantity and quality of work. Quantity and quality were *process* determined. Jobs were very routine and supervision was concerned with adequate training of employees in the various work steps and in the performance of their duties. Even though working conditions were unfavorable, there was very little differential impact on jobs. Limited but meaningful opportunity for informal resolution of problems existed. As will be noted in Chapter 4, the responsibilities of foremen could be programmed almost to the same degree as could employee duties.

In this environment management had established a close relationship with employees and utilized successfully a system of progressive discipline. Union militancy was not a problem. To illustrate the interdependence of variables, it can be pointed out here that at an earlier period in the history of the plant, the coke ovens had been a very high grievance rate department in the presence of inadequate management and militant unionism.

At Company 2 a somewhat different type of comparison was made—that is, steelmaking departments characterized by low grievance rates were compared to wire mill departments which had relatively high grievance rates. A sample of three departments in each category was studied. Variations in task requirements and operating conditions were clearly significant in this contrast, but union and management differences also existed, as will be noted in subsequent chapters. Some differences were also noted among departments within each of these two broad classifications; however, the major contrast in the quantity and quality of grievance activity was captured by the major comparison.

The most important characteristics distinguishing the wire and steel mills were the individual as compared to the team nature of the operations and the difference in the amount of continuous attention the work required. Operations in the steel mills were generally team operations, that is, they were performed by groups of men working together in association with major places of equipment such as open hearth furnaces, rolling mills, rod mills, and so forth. Some operations, such as the rod mills, were continuous and repetitive. Other procedures, such as the open hearth, were batch operations that offered occasional breaks from concerted activity. In the continuous processes, however, spellmen were available so that individuals could take breaks. One immediate result of this technology was that in many situations an individual could work without being in peak physical condition on a particular day because other members of the team could and would pick up the slack. These conditions may well be related to the significantly lower absenteeism rate in the steel mills in comparison to the wire mills; this was reflected in both the use of discipline and related grievance activity. The wire mills, on the other hand, were individual operations which tended to demand continuous attention.

Another reflection of the difference in technology was in the relative wage incentive systems. The jobs in the steel mills were classified according to the general industry plan and the incentive was based on tonnage produced. As such, payment was a function of group effort to maintain high equipment utilization. Each member of the crew was rewarded by the same percentage bonus so that one individual was not pitted against another individual. Also a high proportion of the individual's work was "attention time," which did not require keeping pace with the machine. This was in marked contrast to the individual piece-rate payment in the wire mills. A major difference in grievance activity, as will be noted subsequently, was in the relative number of incentive grievances.

Finally, the steel mills provided much greater opportunity for informal resolution of problems than did the wire mills. Breaks in the production process or the availability of relief workers meant that there was time during the course of a regular day for union grievance committeemen, and employees as well, to meet with supervisors to handle problems or differences informally. In the wire mills, task requirements, augmented by the high noise level, gave no real opportunity for informal discussion, and few were held.

The very low plant grievance rate at Company 4 has already been noted. This rate appeared to reflect the small plant size, favorable community conditions, and the general quality and character of union-management relations. In all departments in this plant informal resolution of problems prevailed, and departmental differences were muted and less marked than in other paired comparisons. The high-rate maintenance department, however, did have characteristics, as was true of two other maintenance departments, which tended to create a modestly high relative grievance rate. This department will be discussed subsequently. The low-rate silicon department also had some favorable operating conditions, but the very low grievance rate, as shown by the record of written grievances, was marred at the time of the study by the use of slowdown tactics which did not find their way into the grievance record. Thus, while the department had a very low grievance rate, the low rate understated the degree of union-management controversy prevailing.

The new silicon department did have favorable operating conditions, comparable to those discussed for all steelmaking departments. The continuous annealing and normalizing lines were highly automated and did not require significant worker involvement for quality and quantity production. The new department and equipment provided more favorable operating conditions than did the old department. Opportunities existed for informal resolution of problems. On the other hand, operating conditions were not markedly more favorable than in other similar departments. What gave the department its especially low rate was the strong desire of the union committeeman for informal resolution, although again the good record must be discounted because of the use of pressure tactics.

Although differences between high- and low-rate departments have been noted in the discussion of low grievance situations, these differences can be sharpened by specific discussion of high-rate departments. At Company 5 the low-rate utilities department was paired with the coating department which coated cold-rolled strip with a variety of materials. A galvanizing line was a major activity in the department. This department also provided a meaningful contrast with the silicon department discussed earlier. The coating process, to maintain quality standards relating to the surface characteristics of the material, required a high degree of worker attention. With this relatively high degree of employee responsibility for product quality, management in turn exercised quite close supervision over employees, including the frequent use of disciplinary penalties. These penalties in turn were the source of frequent grievances.

The high grievance rate in the coating department, which will be discussed in the next two chapters, although clearly related to task requirements and to the aggressive character of union representation, was also associated with a strong policy-oriented management. Although this combination of variables, including the relatively high problem rate created by the task requirements, gave rise to a high grievance rate, the department did not present a highly serious problem from a management point of view. Management policy was being effectively applied.

A second high-rate situation was found in the wire department in Company 2. In these wire mills, operations were basically individual. That is, an individual (or on occasion a helper as well) performed specific operations alone, not in conjunction with others. For example, in the fence department, wire fence was produced on a machine. Each machine was run by a single operator. Time away from the machine reduced production and operator earnings, which generally acted to reduce the amount of informal problem solving done. The continuous attention required by the task, the system of incentive payment, and the high noise level combined to reduce informal problem solving to a very low level.

An important ramification of the difference between the wire and steel mills at Company 2 was evident in the prevailing wage systems, even though dissimilarities in production and attention pressures were in other respects conducive to grievances. The individual piece-rate system in the wire department was replete with alleged inequities in earnings and effort, in part because rates had not been rationalized by job evaluation. Workers in some departments, in part through the use of pressure tactics, had raised their earnings through piece-rate adjustments to levels higher than the work effort merited. These workers and groups of workers tended to use the grievance procedure to maintain their relatively favorable earnings, while other workers with an unfavorable wage rate used the procedure to enhance their position.

The contrast in incentive and other wage grievances between the steel and wire mills—the former with an evaluated structure and group incentive systems and the latter with an unevaluated wage structure and piecework earnings—was marked. (See Exhibit 3.)

Exhibit 3. Breakdown in Grievances

	1964		1965		1966		Fiscal 1967	
	Number	Percent of Total	Number	Percent of Total	Number	Percent of Total	Number	Percent of Total
Department								
Steel Mills								
Incentive	2	6.5	—	—	1	6.7	1	2.2
Other wage	2	6.5	2	6.9	—	—	5	10.9
Total	4	13.0	2	6.9	1	6.7	6	13.1
Wire mill								
Incentive	51	36.7	45	27.3	24	28.2	63	38.2
Other wage	13	9.4	12	7.3	4	4.7	8	4.8
Total	64	46.1	57	34.6	28	32.9	71	43.0

In two of the wire mill departments other environmental factors acted to reduce the opportunities for informal problem solving. In the muffle department the general foreman was in charge of other areas in addition to his department, and thus was responsible for a large physical area. More importantly, part of the muffle department was located in another part of the plant and was not within easy distance of the general foreman's office. A similar condition prevailed in the fence department, in which the general foreman was also responsible for another department not immediately adjacent, and thus he too had to cover a large physical area. As a consequence of all these various circumstances, union-management discussions were scheduled outside the department and were conducted as formal grievance activity.

The remaining two high-grievance-rate departments at Companies 1 and 4 were both maintenance operations and were contrasted with regular production operations: at Company 4 with the silicon department, which finished silicon steel, and at Company 1 with the coke ovens department, which turned coal into coke and assorted by-products. At Company 4 the entire maintenance operation was studied, but at Company 1, because of the large size of the plant, just a small part of the maintenance operations, the bricklayer department, was studied. The qualities of maintenance operations which made them quite different from production operations were evident in both of the maintenance departments.

Like the work done in the utilities department at Company 5, maintenance work was performed at every equipment location in a plant. However, unlike the utilities department, it was not done consistently in every location. That is, the work was not carried out on a day-to-day basis but rather was done when needed or on a scheduled basis. As a result, maintenance work, particularly smaller jobs, was not well supervised. Supervision had to rely largely on the skill and attitude of the employee to assure the correct and swift completion of a job.

Although there were recurring tasks in maintenance work, the probability of a worker on a given day performing exactly the same task that he had performed the day before was light. The job might be similar, but it would not be identical, and it might well be a different chore altogether. Since work assignments varied each day, depending on maintenance needs, there were bound to be "good" and "bad" jobs, and any type of job assignment scheme gave rise to dissatisfaction that was aggravated when incentive pay was involved. Also, differing and changing priorities on maintenance work meant that men were moved from job to job as the priorities changed. Such movement could mean inefficiency and could also cause dissatisfaction on the part of an employee who had taken time to set up a job and then could not complete it.

Another difference between maintenance and production work was that in the former the work of a variety of craft groups might be needed on a single project, and the work of each craft group had to be coordinated so that no one group experienced delay. Also, the work performed by craft groups away from their own maintenance shops was done in a physical area along with production work and was in part under the jurisdiction and at the demand of a separate group of management.

In Company 1, 95 percent of bricklayer work was covered by incentives. Because of the variety of jobs done in the department, and the variations thereon, there were over 1,500 different rates (or standard hours allowed) in use. In addition, the existence of off-standard conditions under which the work frequently was done could cause the alteration of standard rates. Some work also was done under a "duplex" rate system, in which performance below 100 percent of standard still yielded incentive earnings. Under this system, all work above a certain percentage of standard but below 100 percent was paid a fractional bonus. Above 100 percent of standard the regular incentive took over. The complexity of the incentive system used in the bricklayer department and variations in working conditions invited complaints. In 1965 incentive and related wage grievances were 17.6 percent of all grievances filed. In 1966 such grievances accounted for about 30 percent of the total.

In both Company 1 and Company 4 the nature of the maintenance operation made informal problem solving more difficult than in the low grievance production departments. The dispersal of maintenance crews over the entire plant, with only sporadic supervision, was the primary cause of the reduction in the number of opportunities for informal problem solving. To this must be added the demands on maintenance crews, during certain times at least, for quick service to equipment. During these times management priorities placed heavy emphasis on task accomplishment. As a consequence of the combination of task conditions, grievance activity in maintenance operations was almost totally formal in character. In contrast, the production departments, with more regular operations and single locations were much more amenable to informal problem solving. Also, production problems tended to be recurring ones, which did not require new solutions.

A strong contrast was apparent between the craft employees who made up maintenance departments and regular production employees. If any group of employees studied stood out as being different from others, it was maintenance employees. They had a much stronger sense of job ownership than did production workers, and probably a greater sense of pride in workmanship. Therefore, grievances filed under the local working conditions section of the agreement claimed that another craft group or a production group had been permitted to do work which "belonged" to the protesting craft group. For example, in the Company 1 bricklayer department, most of the twenty-four local working condition grievances filed in 1965—20 percent of the total grievances filed—were over jurisdictional disputes. In the Company 4 maintenance department, from 10 to 60 percent of the grievances filed were jurisdictional.

Maintenance employees also showed a greater collective concern for job security than did their counterparts in regular production jobs. In the course of this study, all subcontracting grievances were filed by maintenance employees, whose work, of course, was more likely to be contracted out. However, some of these subcontracting grievances were filed when these same employees were working six- or seven-day weeks. In the Company 4 maintenance department, grievances over job protection, including both jurisdictional disputes and subcontracting complaints, accounted for almost 50 percent of the grievances filed in the period 1964-1967. In a millwright department studied in detail at Company 2, the proportion was 36 percent.

The total percentage of similar grievances was slightly smaller in the Company 1 bricklayer department, about 30 percent, but the absolute number of grievances was greater there. The grievances relating to job protection in the bricklayer department were more easily understandable than they were elsewhere since this department was most threatened by technological changes. However, the grievances over job protection in the bricklayer department cannot be attributed solely to the insecurity caused by technological change, since they were found in other maintenance departments and appear to be the result of the character of maintenance employees and their task requirements and the employment conditions under which they work.

There was a marked difference in the seriousness of the grievance problems in the Company 1 and Company 4 maintenance departments. The bricklayer department at Company 1 had high job insecurity because of technological change and a very difficult civil rights problem associated with militant union leadership. In addition to the very special union situation, management weakness was also involved. Company 4 illustrates maintenance problems less influenced by union militancy and management weakness. It does reveal the tendency of some operating conditions in maintenance to create special problems and formal grievance activity rather than informal problem resolution.

Variations in task requirements and operating conditions clearly give rise to differential problem rates and to conditions more or less favorable to the informal resolution of such problems as do arise. The higher the required attention time, the greater the employee responsibility for the quantity and quality of production, the more variable the task and its location, and the more individual the work assignment the higher the problem rate and the more difficult management's control of production. These factors probably do not capture all the differences in task requirements and operating conditions contributing to high and low formal grievance activity and to the number and character of problems and pressures felt by employees, but there can be little doubt that some operating environments are much freer of problems and pressures and some are more conducive to informal resolution than others. In steel these differences are frequently magnified by incentive payment systems and the resultant grievances.

Summary

The concept of environmental factors—the context within which the union and management operate—is somewhat difficult and complex because the world in which the parties relate to each other involves many diverse things. The concept is made especially difficult because environmental variables may be regarded as both active and passive. The most obvious effect of environmental variables are those associated with economic, technological, social, and other changes which have a direct effect upon employees, typically leading to management action and possible union reaction. On the other hand, when these variables are essentially stable, they play a conditioning role associated with particular environmental characteristics that is very difficult to assess. In a favorable, stable environment a high degree of wisdom and enlightenment may be attributed to the parties, but in reality it may be little more than a reflection of favorable environmental conditions.

Environmental factors were discussed under three broad headings: socioeconomic considerations, technological change, and task organization and the work environment. At the time of the study, general business conditions were favorable and did not significantly differentiate one situation from another. However, size of plant and community environment appeared to be important conditioning influences. The lowest grievance rates were found in a relatively small plant in an essentially agricultural community. The belief was that community characteristics contributed in important ways to the accommodating relationship and the informal resolution of problems was partially facilitated by the smaller size of union and management organizations, the shorter internal lines of communication, and the establishment of relationships off as well as on the job. Civil rights emerged as an important influence but only in some situations. As the discussion of union leadership will illustrate, civil rights problems can at times mean the creation of a Black interest group that cuts across traditional union and management lines and becomes a separate power center.

Technological change can have favorable or unfavorable impacts upon job security just as can changes in business conditions. Technological change can also increase or reduce skill requirements and have other effects upon job content; these in turn may have associated wage and earnings implications. Unfavorable effects from technological change and adverse product market conditions are widely recognized. The study also revealed, however, that contemplated and actual adverse industrial changes alter the *climate* of the relationship as well as serving as a direct source of grievances. Neither technological change nor socioeconomic variables consistently distinguished high-grievance-rate departments from low-grievance-rate departments.

The most interesting environmental variable centered around differences in task requirements and conditions, which in turn were created by differences in technical and product market conditions. In part these dissimilarities were a matter of economic and technical stability and in part a matter of differences in production processes. What was clear was that routine operating conditions created far more day-to-day problems in some departments than in others. What was also of at least equal significance was that working conditions in some departments were far more conducive to informal resolution of problems than in others. It was differences in what might be called the normal operating environment that had the most notable impact upon the creation and resolution of day-to-day problems. It was obvious, for example, that incentive problems were more difficult in some departments. Special problems, such as job assignment in maintenance work, were associated with particular production conditions. In short, some work conditions and task requirements were decidedly conducive to a low or high grievance rate.

The discussion of union and management variables which follows should thus be considered within the context of the preceding discussion. Management and union organizations and individuals, operating in a particular environment, relate economic, social, and technical forces and conditions to specific problems in the work place. Neither an environmental nor a behavioral analysis of grievance challenge and resolution alone appears to provide an adequate explanation of the process. A favorable environment and a low problem rate do not assure a low grievance rate.

3 Union Influences on the Challenge Rate and the Resolution Process

IN THE COMPANIES AND PLANTS SELECTED for this study union representatives played a major role in the resolution process. Some employees were willing to discuss a complaint or problem with a foreman or general foreman, but many others would do so only with a union representative present. Union representatives were therefore frequently involved in the resolution process from the earliest informal stage, if only because the individuals willing to see a supervisor alone might well see a union representative first. When a complaint was put in writing, it always involved the union.

Three aspects of the union impact on the challenge rate and the problem-solving process will be considered here: union leadership, union organization, and union policy. The section on union leadership will discuss the nature and quality of the leadership and its impact on the challenge rate and the process, as well as the relationship of leadership to internal union politics. The section on union organization will define the locus and impact of decision-making authority, and the section on policy will briefly discuss particular policies pursued by some local union groups and their impact on the formal challenge rate and the resolution process.

Union Leadership

Union leadership is the single most conspicuous and outstanding variable in the generation and resolution of problems. The character and type of union leadership largely determines whether the type of union challenge is militant or nonmilitant. The challenge rate is in part conditioned by the fact that unions are political organizations and vary in the degree of their internal political stability. Also, the union committeeman, an elected official with his own constituency, can be highly independent at the department level.

The nature of the union organization tends to focus authority on one or two individuals in the union hierarchy for grievance handling. These individuals are key union people involved in the resolution process and their leadership styles are crucial for that task.

Five different types of union leadership were identified during the course of the study; or to put it another way, five basic patterns of union leadership behavior were observed. These types have been labeled: *inactive, problem solver, advocate, politician,* and *radical.* The inactive and the problem solver are associated with a low grievance rate; the other three are associated with a high grievance rate. The leadership styles, or behavior patterns will be described in some detail. But it should be noted that the study in the steel industry did not produce any perfect examples of each of the leadership types described. That is, the "pure" types did not exist in real life. For example, an individual might largely be a problem solver, but he also was required to have a bit of the politician in his makeup to survive as a union leader.

Inactive leadership produces very few challenges, formal or informal, to management decisions. Inactive leadership has relatively little formal contact with supervision. Leadership which was in essence inactive was found only in one location during this study: the Company 1 coke ovens department. But union leadership here was not completely inactive. Grievances were filed, and the grievance committeemen did have some discussions with supervision. However, many of the foremen interviewed did not even know who the union representatives were, and in this department the foremen played a strong role in the employee problem-resolution process. The strong position played by the foremen probably acted to reduce the role of the union leadership, as did a stable environment that did not generate many problems.

The use of the word "inactive" to describe union leadership does *not* imply that employees were adversely affected by the actions (or inaction) of that leadership. This type of leadership was found in a department where management had good knowledge of the labor agreement and took great care to see that its provisions were followed. Management ran a tight ship, but not so far as could be determined at the expense of any employee rights under the agreement. Thus in a very real sense the limited union challenge in this department appeared to be consistent with the requirements of the situation.

Nevertheless, the grievance rate in the coke ovens department was extremely low, and the only management decisions that were challenged regularly, either formally or informally were those concerning discipline. However, the well-developed and established progressive discipline system in use made such challenges difficult. Disciplinary actions, particularly discharge, affected the employee involved so much that it was difficult for the union to avoid grieving them, particularly under the constraints imposed by the Landrum-Griffin Act, even though the union officials admitted privately that in most instances there was little or inadequate substance to their cases.

Some individuals in the union at Company 5, as distinct from Company 1 above, considered the international representative who serviced the local there to be an inactive leader. Their feelings were based on the relatively small number of cases that this representative carried to arbitration as compared to his predecessor. Based on this quantitative measure, he was less active. Again, this did not necessarily mean that the membership suffered from his actions, since this individual preferred to expend a great deal of energy toward informal, high-level resolution before (and really in place of) arbitration. His challenges to management action, while part of the formal grievance procedure, were still informal. Thus, while in one sense this individual might be considered inactive, in another he would have to be called a decidedly active problem solver.

The *problem-solver* type of leadership was found in all of the low-grievance-rate departments studied, with the exception of that at Company 1 which was discussed earlier. The key attributes of the problem solver were a willingness to compromise, a preference for informal challenge and resolution, a selective approach to worker complaints, and a considerable willingness to be guided by precedent. The problem solver appeared to desire a socially stable environment through the establishment of a judicial framework.

Sayles and Strauss note that screening out some complaints helped to make the position of the grievance committeeman stronger in those issues that he did present.[1] Thus, the problem solver could be an effective union leader because he tended to devote his time to those problems which were important. Also, since he was somewhat selective in the problems he brought for grievance resolution, and reasonable as well, management was more receptive and responsive to such an individual than to other types of union leadership.

Union leaders who could be classified as problem solvers were found at the department level in both the Company 3 structural mills studied. The Plant A structural mill had two grievance committeemen, one of whom was a problem solver; the other, a close associate of a second local official, who will be discussed later, was not. The grievance committeeman in the utilities department at Company 5 was perhaps a classic example of a problem solver in his relationship with supervision and in his practice of keeping all complaints in the oral stage until time limits specified in the contract forced him to reduce them to writing. At Company 4, the individual who held the office of grievance committeeman in the silicon department prior to the time of this study was also a problem solver, and he boasted (not quite correctly) that in three years on the job he never found it necessary to file a grievance. His successor, al-

---

[1] Leonard R. Sayles and George Strauss, *The Local Union* (rev. ed.; New York: Harcourt, Brace and World, 1967), p. 47.

though fairly new on the job, seemed to be following a similar course.

As a further, more specific example of a problem-solver leadership style, an incident in the steel mills division at Company 2 can be cited. Two men claimed to have missed an opportunity for working overtime because of a supervisory scheduling error. Both approached a grievance committeeman with their complaint. The committeeman told one of the men that his grievance was not valid and refused to pursue it further. He mentioned the other complaint to the superintendent, who agreed that a mistake had been made. The superintendent offered to allow the man to make up the overtime as soon as possible, a solution to which the committeeman agreed.

In many places observed in this study, including other departments in this particular plant, the standard solution to this type of problem was to pay the employee for the overtime missed but not worked. This standard solution was not mentioned in this case, and the committeeman found the superintendent's proposal acceptable. Indeed, the solution did remedy the wrong, but at no additional cost to the company. The committeeman might have taken both complaints to the superintendent and bargained a bit. He might have filed a written complaint. He did neither of these things, thus providing a good example of the problem-solver type of union leadership at work. The supervisory response, however, was critical in this instance. The management side of the equation will be discussed in the next chapter.

In departments where the union representatives were problem solvers, the formal grievance rate was probably slightly higher than in departments where the leadership was inactive. Informal activity, however, is inadequately reflected in the rate. Also, interactions tended to be more two-sided, and the resolution of complaints or grievances was really essentially in the interest of both union and management. Problem solvers faced the same kind of political problems that inactive leadership did with disciplinary penalties. Seniority issues were also difficult, and some of them tended to get into the grievance procedure simply because any decision by the committeeman would frequently alienate some group or individual. A committeeman could not always be a "pure" problem-solving type.

The hypothesis could be advanced that favorable task requirements and surrounding conditions were conducive to somewhat inactive or problem-solving leadership. The incidence of inactive leadership was too small to be given weight. It would be very interesting to test the relationship of environment to problem-solving leadership in a larger sample. The evidence here can do no more than suggest a possible positive relationship, as will be indicated in the summary to this chapter.

The *advocate,* the third type of union leadership identified, took what might be called a lawyer's attitude toward his responsibilities and had a strong loyalty to the union institution. The advocate generally assumed that his function was to represent the employees who had elected him to office, without regard to the "equity" of their complaints. That is, no matter what the complaint, the advocate pressed the employee point of view as forcefully as possible and to the highest conceivable step of the grievance procedure. Typically, the advocate was unwilling to compromise an issue and was reluctant to refuse to handle or to drop a case.

The grievance committeeman in the maintenance department at Company 4 tended to take an advocate's point of view, and, in addition had a forceful (and to management, distasteful) way of expressing himself. However, the best example of an advocate was found in the individual who held the position of chairman of the grievance committee of the maintenance local at Company 1. The results of his activities were really quite amazing. The workload at the upper steps of the grievance procedure at the plant studied was extremely heavy, with over 1,200 cases waiting to be heard at Step 3 at the time of the study. This heavy load was primarily due to the maintenance local, whose appeals to that level comprised an appreciable part of the *corporate* total. This individual took an extremely large number of cases to arbitration, including cases which were identical to others that had previously been decided by the permanent umpire. The local had gone from a state of relative prosperity to a hand-to-mouth existence bordering on virtual bankruptcy because of their inordinately high arbitration expenses.

The attitude taken by those union leaders who assumed the advocate's role was understandable. It was easy to see why a union representative might choose to behave in this manner. He could never be accused of not doing his job, he never had to turn away a fellow employee, and it was relatively easy for him to get re-elected. He was a very strong unionist who generally could be said to believe that management was always wrong or was always attempting to "put one over" on the unsuspecting employee. However, the nature of any resolution process does not lend itself to the absolute primacy of any one viewpoint, but rather to balanced decisions. Also, management did not respond too constructively to a union representative who took the point of view that management's actions were always wrong. Thus, where the union representative assumed the role of advocate, the grievance rate was higher than it was under a problem solver, and informal resolution was impaired because of the dispositions of both the union and the management. It is not possible to measure the relative "real" accomplishment of the two types.

The *politician*, the fourth leadership type, was found in only two instances, and neither one at the departmental level, but rather higher in the union organization. To some extent the activity of advocate and radical leadership was political in that it was probably easier for such leaders to get re-elected than problem solvers, and their actions were at least in part based on this knowledge. However, the individual identified as a politician was not interested in problem solving or advocacy as such, but used the grievance procedure to preserve and advance his political position. Thus, the politician, like the advocate, handled all cases, but unlike the advocate did not take cases as far as he might, that is, to Step 4 or arbitration. The politician used formal challenges and the natural delays as well as self-created delays in the grievance procedure to dim the memory of the issue and to give the impression of great effort expended on behalf of the employee. He was not necessarily interested in seeing the union point of view on particular issues vindicated, as was the advocate. The politician also tended to handle problems in a highly visible manner outside the regular grievance procedure, employing appeals to the National Labor Relations Board, and federal and state labor departments, and often accompanying his actions by somewhat inflammatory statements to the news media. Additional evidence of the political motivations of this leader has not been included in this report, but only one other union leader observed in this study clearly was heavily motivated by acting in a manner best calculated to enhance his political future.

The *radical* was similar to the politician in that he used

outside agencies and had basic interests and motivations other than simple problem solving. However, the radical, or "man-with-a-cause," used the grievance procedure not for personal advancement, but to serve a cause toward which he felt strong loyalty. The chairman of the grievance committee of one local union appeared to be a political radical, interested more in creating conflict than in resolving problems. However, he was politically isolated within the local and as a result was not so successful as he might otherwise have been.

The grievance committeeman for the bricklayer laborers at Company 1 was much more successful in his cause, the advancement of civil rights; at least he was successful in airing the problem he thought existed. In addition to being the representative of the bricklayer laborers, he was the vice-president of the maintenance local, the plant civil rights chairman, and an active member of the executive board of the area's Urban League. As a Black he was chiefly interested in improving the lot of his fellow Blacks both in and beyond his department. In his role as department grievance committeeman, many of the grievances he filed contained charges of racial discrimination or had charges of discrimination as a basis for complaint. His grievances also were designed to protect and enhance the jobs and status of his constituents, more as Blacks than as regular employees. To the observer, his ability to verbally abuse mainly White supervisors without penalty enhanced his position with his constituency and helped him maintain his leadership position. He also used appeals to outside agencies to attempt to achieve his ends.

The use of the grievance procedure to pursue civil rights goals resulted in a grievance rate higher than would ordinarily be expected in a situation like the bricklayer department. The combination of the activities of this grievance committeeman with those of the chairman of the grievance committee of the maintenance local created an extremely difficult situation for management. The management response to the grievance committeeman was to attempt to treat and resolve the difficulties that he developed as regular labor relations problems through the grievance procedure. The problems the committeeman brought up did not really belong in the grievance procedure in the sense that the contract was being continuously violated by management and the cases were so complex that they could not potentially be solved through the regular procedures at all.

The union leadership in the coating department at Company 5 was not mentioned in the preceding discussion. The grievance committeeman in the coating department at Company 5 was difficult to classify, mainly because his relationship and his behavior differed toward different members of supervision. His behavior ranged from advocate to radical (the latter, at least, in the eyes of some supervisors) and was not what could be called problem solving.

Union authority at Company 2 was extremely diffused through a committee structure. Union committee members in the low-grievance-rate steel mills departments tended to be more reasonable and less militant and had a much more definite problem-solving bent than did their counterparts in the wire mill. The example used earlier in the chapter is indicative of their approach.

All of the high-grievance-rate departments studied had union representatives, either at the department or local union level, who were advocates, politicians, or radicals, while all of the low-grievance-rate departments had representatives who were inactive or problem solvers. In Company 1, the coke ovens department leadership was inactive and the local leadership moderate with a general problem solving, while the outlook of wire mills leadership had radical leadership and the chairman of the local's grievance committee was a strong advocate type. At Company 2, the steel mill's leadership was oriented toward problem-solving, while the outlook of wire mills leadership leaned toward the advocate type. In one plant of Company 3, the leadership in the structural mill was at least partially problem solving in nature, but was counteracted in part by strongly political local union leadership. In contrast, the leadership in both the structural mill and the local union in the second plant were problem solvers. At Company 4, the grievance committeeman in the high-rate maintenance department was generally an advocate, while the lower-rate silicon department had a committeeman who could be classified as a problem solver. Finally, at Company 5, the coating department grievance committeeman, though somewhat ambiguous, still could definitely be classified as at least an advocate, while his counterpart in the utilities department was classified as a problem solver.

From the preceding discussion it should be obvious that although "pure" types of union leadership were described, the examples used as illustrations were not able to be precisely categorized. On the other hand, these types of leadership were not invented prior to the research, nor were they devised to fit different levels of grievance activity. It became clear as the study progressed that not only did departmental challenge and resolution tend to be concentrated in individual committeemen, but that these individuals had different leadership styles, motivations, and patterns of behavior at least in terms of their approach to grievance challenge. Although particular types of motivations and loyalties tended to be associated with higher or lower levels of formal grievance activity, each type in a large study would no doubt be associated with a considerable range of grievance activity. But it seems quite clear that the differences in motivation and behavior are significant.

Leadership behavior also reflected the internal political situation in the local union. Unions are political organizations, and all union officers and committeemen must campaign if they are to remain in office. Unions also tend to factionalize, and such factionalization might be expected to affect the grievance procedure as different groups use the union in an attempt to advance their own position. The year 1967 was an election year for local union officials, and in every location studied the grievance rate for that year was higher than in previous years. This increase was widely attributed by company officials to election-year politicking.

Specific examples of the effects of internal union politics were somewhat difficult to find, at least consistently. At Plant A of Company 3, twenty-odd seniority grievances from a particular structural mill group, which had been filed before the union election in 1967, were still pending in the procedure at the time of the study. The local president said that if he were elected he would press the seniority issue in them through the grievance procedure, while his opponent made no statement about them, but told management privately that they were without merit and would be dropped if he were elected. Management believed that the cases were without merit, and their continued presence in the procedure was a function of union politics even if their initial injection into the procedure was not.

At Plant B of Company 3, all the officers and nine of the

ten grievance committeemen were new and had won their positions during the 1967 elections as a result of dissatisfaction on the part of the membership with the previous administration. The winners waged a militant campaign and management feared that they might be difficult to deal with, but such had not been the case. The issue in the election was specific mistakes by the incumbent administration, most specifically a seniority issue, and there was no allegation of a general failure to deal with management effectively. This fact may have explained the subsequent moderation of the new officials and the weak election reaction that appeared only as a slight increase in the grievance rate.

At Company 2 the union was probably more factionalized than at any other location visited. However, the typical disruptive effects of competitive factions were difficult to see on the department level, in part the result of the committee structure. Three factions could be identified. One, centering around the incumbent officers, particularly the president, was bent on a militant course of action, including the expansion of arbitration activities. Another faction was simply less militant and wanted a return to a more harmonious labor-management relationship. The third faction wanted the independent union at Company 2 to affiliate with the United Steelworkers in order to be more successful with management. The factions existed from the time of a strike in 1966, though since that time the factionalism had decreased somewhat.

In one part of the open hearth department at Company 2, the stocker section, factionalism appeared to affect the grievance rate. Older and younger employees were in contention over the employee assignment to certain lower-paying jobs. The older employees wanted the jobs to be assigned, as they had been, on the basis of seniority, which meant that the younger employees received the poorer assignments. The younger employees wanted the jobs rotated among all employees. Several grievances were filed as supervision changed from seniority-based assignment to rotation and back again at the request of these employee groups. Supervision had no preference in the matter, but was caught in the crossfire between two groups. The issue was finally settled by an election among the employees.

At Company 5 the local union was also somewhat factionalized, and the 1967 elections saw one group take all the top offices except one. The lone holdover was the chairman of the grievance committee, and his politics on the local level as well as those in a more general sense helped isolate him within the local; as a result, his impact on the resolution procedure was blunted to a great extent. Again, a slight increase in the grievance rate was noticeable at Company 5, in 1967, but clear, specific examples of the results of factionalism were not apparent. However, in this case and in the others, the rise in the grievance rate in 1967 indicated that almost all committeemen evidenced a need to make their activities on the behalf of constituents visible during an election year and that they could not afford to alienate constituents by failing to handle complaints. Thus union politics do influence the grievance rate, particularly in an election year. In a study comparing departments, behavioral differences among union leaders were more significant than broader political differences. A study comparing plants might be expected to reveal more significant differences in the level of political activity, but the suspicion is that individual leadership behavior would remain decidedly important.

Union Organization

Union organization generally paralleled that of management. Each department studied had one or two grievance committeemen, corresponding to middle and top department management. Each committeeman had under his jurisdiction one or more stewards, who corresponded to the foremen. In theory the stewards were supposed to represent employees in discussions with foremen, but in practice the stewards, like the foremen, had little or no role in the resolution procedure. The steward was part of the union political machinery and not a significant part of the grievance procedure. This phenomenon, though evident in the steel industry, is not limited to it, as Sayles and Strauss and others have indicated.[2] In the steel industry the role of the foreman is generally minimal in terms of the grievance procedure, as will be seen in the next chapter.

The grievance committeeman, then, was the chief union representative at the department level. He dealt occasionally with foremen and much more often with general foremen and the superintendent. Because the roles of both the stewards and the foremen were in practice greatly diminished (compared to what they theoretically might have been), a tendency existed toward the contraction of the grievance procedure outlined in the labor agreement. Problems were frequently taken to the superintendent's level informally, and even beyond, often involving the labor relations staff. As a result, the second and even the third steps of the formal written grievance procedure

---

[2]*Ibid.*, pp. 16–20. See also: *The Grievance Process,* Labor and Industrial Relations Center, Michigan State University, 1956. Several participants at this conference noted the particular role of the steward, but see especially William F. White, "The Grievance Procedure and Plant Society," pp. 14 ff.

often were carried out as a formality to record informal understandings or only after informal procedure had not brought resolution.

The company was represented in the third step hearings by the labor relations department. The production department's grievance committeeman was usually present, but the union's case was handled by the chairman of the grievance committee. At Company 3 and Company 4 the chairman of the grievance committee was also the president of the local union. At Company 1 and Company 5 the grievance committee chairmanship was a separate office. The union organization at Company 2 was unique, and it will be discussed separately.

The departmental grievance committeeman was responsible for appealing grievances to Step 3. After they were appealed, cases became the responsibility of the grievance committee chairman. Occasionally a grievance would be appealed to Step 3 and then withdrawn without an actual hearing. At the Step 3 level, before and after the actual hearing, the entire grievance committee acted, at least in name, as a screening committee. This was true everywhere except the maintenance local at Company 1, to which the bricklayer department employees belonged. The local had instituted a screening committee just prior to the beginning of the study there, but it was not functioning effectively at the time the study at Company 1 was completed. In some cases, most particularly at Plant A of Company 3 and at Company 5, the grievance screening committee was dominated by its chairman, who, as will be noted later, used the grievance procedure for purposes apart from the resolution of problems. Even in these circumstances, in addition to the more normal situations, the screening committee was a device used by departmental grievance committeemen to avoid the political consequences of refusing to handle weak cases. The committeeman, rather than refusing to handle a complaint, would file a grievance and then drop it at or after Step 3, telling the constituent that it had been screened out by the committee. The nonexistence of a screening committee in the maintenance local at Company 1 led to an increased grievance load at Steps 3, 4, and 5. The nature of the leadership provided by the chairman of the grievance committee, who in the absence of a screening committee was responsible for the situation, will be discussed shortly.

The union's case at Steps 4 and 5 (the latter being arbitration) was handled by a staff representative from the international union appointed by the district director to service the local union. Generally speaking, the international representative was required to dispose of cases as instructed by the local union. However, in some instances, the personal influence of the international representative, who frequently was a former officer of the local, gave him a somewhat independent role in the resolution process. This was true at Plant B of Company 3, at Company 4, and in one local union at Company 1. In these instances the international representative's influence was more personal than positional, but in any event was substantial. At Company 5 the staff representative's independence dated from an earlier period when the local was held in trusteeship by the international. As a result, his authority was more positional than personal, but he still had a great deal of autonomy at the fourth and fifth steps of the procedure. At Plant A of Company 3 and in the maintenance local at Company 1, the international representative had little personal influence; as a result, more cases tended to be taken to Steps 4 and 5. This was much more evident at Company 1 than in the other locations where staff representatives used their influence to keep all but important cases from the upper steps. The international representatives tended to have better ability to judge the chances of winning cases in arbitration.

One impact of the division of a plant's work force into more than one local union was noted in Chapter 2, where the division of the work force at Plant A into three locals meant that the impact of layoffs in the open hearth was not felt in the structural mill, in contrast with Plant B. Because the work force in Plant B was organized into one local union, seniority problems became a matter of internal union concern there, a situation that did not arise in the structural mill local at Plant A. Also, the two departments studied were not only under the direction of different local leaders but different district directors as well, which acted to differentiate the grievance rate and also created variations in the quality of the grievance procedure.

The multiunion arrangement at Company 1 meant that the departments studied also were under the jurisdiction of different local leaders. (The lack of a screening committee in the maintenance local has already been mentioned.) In direct contrast, the local to which coke ovens department employees belonged had used a screening committee for a number of years. The same international representative was assigned to both locals, however. This individual was a former president of the coke ovens local and for that reason may have had more influence over that local than over the second (maintenance, including the bricklayer department), but his efforts had achieved the beginnings of a grievance screening process in the maintenance local. Given the character of the leadership of that local, particularly in the person of the chairman of the grievance committee, this was quite an accomplishment.

The local union at Company 2 was an independent union, not related to the Steelworkers or the AFL-CIO. Its organization was different from any of the other local unions and the resolution process at Company 2 was correspondingly different. At the department level, employees were represented by a committee, usually consisting of three individuals. The person who received the largest number of votes was the chairman, but, generally speaking, it was necessary for supervision to work with the committee as a whole. Working with three individuals rather than one was difficult, particularly since the refusal of one of the committee members to agree to a solution at times resulted in moving a dispute to a higher level. Resolution was more expensive for management, too, since discussions meant that three men rather than one were away from their jobs. Only at Company 2 were union representatives paid by the company, rather than the union, for the time spent in the formal grievance procedure and other meetings away from the workplaces.

Above the department level even more individuals became involved in the resolution process. At the second step of the four-step procedure, the department committee met with either the wire mill or steel mills superintendent. At this level they could be joined by members of the wages and rules committee or the grievance committee, or both. These two five-man committees always operated at Step 3, although not always together. The local's officers were also present at Step 3, where the labor relations department represented management. A contractual limit of fifteen paid union representatives was placed on Step 3 meetings, but again problem resolution was a much more difficult task at this step at Company 2 than elsewhere since it involved so many people.

The diffusion of authority within the union and the accompanying increase in democracy were probably hallmarks of an independent union and not totally unique to Company 2. Although democracy may be increased in and by the procedures used at Company 2, the resolution process was much more difficult and more time-consuming for management. To some degree the procedure may have secured a greater degree of consensus than at other companies, but this is not certain.

The local union at Company 2 really had no locus of power, either formally or informally. The local was an extremely democratic organization, but none of the officers actually seemed to trust one another. In the other locals studied, a distinct locus of power existed, usually in the person of one man. In the maintenance local at Company 1, the key man was the chairman of the grievance committee and president of the local, as might be expected from looking at the formal organization. The power of the chairman of the grievance committee was such that the office was frequently combined with that of president. In yet another variation the extremely militant chairman of the grievance committee at Company 5 was generally undercut by officials at a higher level, including the international representative. The situation was possible because of the power the international representative customarily exercised. In summary, a locus of power generally existed within the local union, was usually vested in one man, and did not necessarily follow the pattern indicated by the formal organization.

If the study were expanded, more differences would

probably be found in the formal union organization and also in the general locus of power. For example, in the construction trades, the local business agent is generally an extremely important individual. As more diverse situations were studied, however, the variable of union organization would be expected to be more and more a function of environment and policy.

The combination of leadership style and center of influence in the union hierarchy had important ramifications for the grievance procedure. First, real authority and autonomy in the resolution process was limited to one or two individuals on the union side. This fact tended to magnify the importance of leadership style as a determinant of the grievance rate and of the resolution process. Also, the presence of certain types of union leadership—advocates, politicians or radicals—at the department level resulted in higher initial grievance rates than with other leadership types. Where high-grievance-rate departmental leadership styles were replicated at higher levels in the union hierarchy, the workload in the higher steps of the grievance procedure, including arbitration, was much greater. This situation presented management with a much more difficult task than would otherwise have been the case.

Two examples are provided by Company 1 and Company 5. In the bricklayer department at Company 1, radical leadership at the departmental level was paired with a strong advocate leadership style on the part of the grievance committee chairman. The result was an almost unmanageable case load for the industrial relations staff at the higher steps of the grievance procedure. As a result they had little time to spend on grievance-prevention activities at the departmental level. Of course, this meant that the initial grievance rate remained high.

At Company 5 the departmental committeeman was an advocate, while the chairman of the grievance committee was a radical. The fact that the grievance committee chairman was politically isolated within the local helped to reduce the impact of his behavior. Also in this case, strong departmental management and a different line-staff relationship meant a somewhat reduced initial rate. Nevertheless, the grievance rate was high and more cases reached the upper stages of the grievance procedure than would have been the case had different union leadership styles prevailed in the two positions in the union hierarchy.

At one company a political leader of the local was coupled with problem-solving leadership at the departmental level, thus reducing the number of problems that came within the jurisdiction of the local president. In this instance, resolution of problems with the local president was quite difficult, so the fewer the grievances that reached the third step in the grievance procedure, the easier was management's resolution task.

Union Policy

In some instances the policy of local unions to pursue certain issues through the grievance procedure had a definite effect on both the grievance rate and the resolution process, at least in terms of the level to which these issues were taken.

The local for the coke ovens at Company 1 followed a policy of grieving displacements caused by every technological change. If in the course of the grievance procedure the changes were not reversed, the issue was taken to arbitration. The one coke ovens case of record taken to arbitration involved displacements (but not layoffs) due to minor technological change. The change was carefully introduced by management with both pre- and post-change studies made by the industrial engineering department. As a result, the new assignments in the work force made as a result of the new technology were sustained by the umpire. Despite the lack of a strong case, the question was arbitrated by the union as a matter of policy, and the fact that the union did pursue this course may have been responsible for the care that management took in the introduction.

The local union at Plant B of Company 3 followed a policy of grieving the description and classification of all new or changed jobs. These grievances were rarely, if ever, settled at the plant level. These cases went to Step 4 level at corporate headquarters where they were settled. Cases that involved no essential conflict were still taken to Step 4, simply as a matter of policy and procedure. Both sides seemed satisfied with the resolution procedure and level as it pertained to job description and classification, and, in the general absence of conflict, this satisfaction must be viewed as the reason for the continuance of this policy by the union.

The union committee in the millwright department in the wire mill at Company 2 followed a policy of filing grievances for all jobs involving millwright work that were contracted out by management. The committee followed this policy when the department was already working at capacity, with most employees working a six-day week or more. This policy was pursued to maintain a claim on work not currently being done by the department because of a capacity constraint. The committee believed that their chances of keeping the work from being contracted out in the future—when the present workload was reduced—was enhanced by this grievance policy. In many ways they were acting to preserve future overtime opportunities, and in answer to their grievances, management assured them that the work which was currently being contracted out "belonged" to the department. This work would be assigned to the regular employees when the workload would permit it.

This policy was mentioned earlier in discussing the concern of maintenance employees with job protection. In the case of the Company 2 millwright department, however, this concern was reflected more in a deliberate or conscious policy than it was in the maintenance departments studied at Company 1 and at Company 4.

As a final note in this discussion, it should be clear that although union leadership, organization, and policy have been discussed singly and somewhat abstractly, they are related in a complex manner to historical events and to the environment. The situation and individuals discussed and described did not "just happen," and did not exist in a vacuum. They had a long history behind them, one that time and space prevented from being investigated and reported.

Summary

Factors relating to the union influenced the *rate* of challenge and its *character*. These union influences were discussed under three headings: union organization, union leadership, and union policy. Differences in union leadership were the most conspicuous variable in the study.

Five types of union leadership were discussed: inactive, the problem solver, the advocate, the politician, and the radical (man with a cause). The first type was found only in a stable setting and in the presence of a strong management that followed the agreement carefully. Thus the necessity and opportunity for challenge to management decisions were somewhat lacking. The problem solver tended to challenge management actions informally and as a consequence the resolution process was informal and the formal grievance rate low. The problem solver also screened out complaints, which helped reduce the grievance rate. All of the low-grievance-rate departments were associated with these first two types of union leaders. The grievance rate was higher with the other types of leadership. Formal challenges to management action without strong regard for merit suited the temperament of the advocate. Such challenge suited the objectives of the politician and the radical who had aims other than the resolution of grievance problems.

Union organization paralleled that of management and influenced the level of settlement and also the challenge rate. Union stewards had no role in the formal or informal resolution process. The major union role at the department level was played by the grievance committeemen. The use of screening committees helped to reduce the number of grievances in the upper steps of the grievance procedure but also may have prompted the formalization of challenges since the committeeman involved could use the screening committee to block the upward movement of a grievance without personal repercussions. The chairman of the grievance committee had a strong role in the formal grievance procedure. The role of the international representative varied. Union politics also influenced the challenge rate in some instances, and this was particularly evident in the increase of the grievance rate in the election year of 1967.

Specific policies pursued by some union groups meant that certain issues were always grieved and frequently taken to a specific level of the formal procedure, without regard to the predisposition of individual committeemen to handle problems informally. The particular policies followed by local unions, while significant, appeared somewhat less important than those of management, which will be discussed in the next chapter. Variations in union leadership were very strongly associated with differences in the grievance rate and in the character and formality of the resolution process. Whether the type of union leadership can in turn be associated with differences in the environment and in management policy must be left as an open question so far as this study is concerned although the interrelationships are obvious.

ized within
4 Management Influences on the Grievance Rate and Resolution Process

ALTHOUGH MANAGEMENT INFLUENCES on grievance activity and the resolution process will be analyzed within a framework identical to that used for the union variables —organization, leadership, and policies—there are substantial differences in the role of the two parties. In an immediate and direct sense management responds to the union grievance challenge as well as to pressures in the external environment. While the management response may demonstrate varying degrees of leadership and initiative, policy orientation, effectiveness, acceptance of the union, and may vary in other respects as well, it remains, particularly in the short run, a response to union-initiated activity. The character of the managerial response, however, affects the grievance rate and the resolution process in highly significant respects.

Variations in the managerial response are less conspicuous and more difficult to identify than are variations in departmental patterns of union activity. There is no highly conspicuous variable, such as was true of union leadership, and, in fact, the importance of variations in managerial leadership behavior is diminished by the diffusion of decision-making power within the management ranks and by policy and procedure constraints. Management leadership variations are also in part responses to status, position, and union pressures, rather than being attributable to dissimilarities in personal leadership styles. Organizational and policy differences are thus relatively more important than leadership in the management analysis even though leadership differences are interesting and significant.

The difficulty in identifying differences in the management response stems from important additional considerations. In the first place, the more significant organizational and policy variations are *corporate* in character and are not directly reflected in differences in *departmental* behavior even though departmental consequences flow from them. In the second place, differences in the management response are not dramatic because of the uniformities in corporate behavior. There are many similarities in the substance and process of the management response which arise from the highly centralized collective bargaining system in basic steel—from almost identical labor agreements, from close association among management executives and their considerable desire for consistent policies, and from similar problems and environmental pressures. Finally, variations in the managerial response are difficult to identify because at the departmental level they are frequently unplanned; and to some extent these unintended differences arise from yielding to particular union pressures or result from unplanned response to variations in the character of the union challenge.

Differences in management organization will be discussed in the first section since variations in line-staff relationships and behavior may well be the most significant of the management variables. Also important as an organizational consideration is the "fit" between line management's organizational structure and locus of authority in relation to differences in the difficulty of the managerial task in various departments. A second section will deal with line management leadership variations and their impact on the grievance rate and the resolution process. A third section on managerial policy will explore some selected policy differences on grievance activity. In particular, attention will be given to policy differences in the areas of discipline, wage incentives, and consultation with the union. These particular policy differences stand out as having important consequences for departmental grievance activity and the resolution process.

Management Organization

• The most conspicuous fact about management's role in the grievance-resolution process is that authority for resolution is shared, in fact if not in theory, by two different management groups: production or line management and industrial relations staff.• The manner in which the resolution task is allocated between these two groups will be shown to affect the grievance rate and the level at which grievances are resolved.• In general, the greater the involvement of industrial relations staff at the earliest stage of the resolution process, the lower the formal grievance rate, the greater the likelihood of informal resolution, and the lower the step at which settlement will occur. To some extent these consequences developed because the final locus of authority for the resolution of grievances in all of the companies rested with the staff, and this was made evident at the third and higher steps of the procedure if not at earlier steps.

Each of the companies, except Company 2, had a distinctive manner and policy of sharing authority, which prevailed in both the high and low grievance departments. Company 2 showed much variation between its steel and wire departments, which amounted to a distinction between plants, in a manner comparable to other corporate differences. Though some of the corporate differences are more significant than others, the somewhat subtle nature of the differences makes it appropriate to discuss each in turn and to relate the perceived consequences in terms of differences in departmental grievance activity and the resolution process.

The manner in which the labor relations staff operated at Company 4 gave it a decidedly different role from that which existed at any of the other companies. First, however, it should be recalled that the grievance rates at Company 4 were very low, much lower than at any of the other companies. The low rates in both departments were a more distinctive feature of grievance activity at Company 4 than the difference between the high and low departments even though these latter differences were similar in character to variations at the other companies. Second, a very high degree of union-management accommodation existed at Company 4—probably higher than at any of the other companies. This accommodation, though a deliberate top-level management objective, was facilitated, as was the grievance resolution process, by the small size of the plant and its location in a rural community.

The low grievance rates at Company 4 were a product

of what appeared to be a relatively low corporate challenge rate (though this cannot be demonstrated), and also to the use of oral resolution for most of the grievances. Since the plant was small, it had only six committeemen. The distinctive feature of grievance resolution was that these committeemen had direct access to the labor relations department for handling complaints and resolving problems. At all other plants union representatives had access to labor relations personnel, below the third step, only indirectly through line management. To be sure, direct contact was made at times in other companies, but union representatives were either directed to department management or were put off until the labor relations representative could confer with department management. Direct labor relations contacts were much less frequent in plants other than at Company 4, where decisions were sometimes made by labor relations staff without consulting line management. Line management was quite dissatisfied with the lack of consultation. They had no more quarrel with the actual decisions made than did line management elsewhere, but they resented their limited involvement in the decision process.

The grievance process as it operated at Company 4 was contrary to textbook prescriptions and clearly had a high degree of shared authority vested in the staff. However, putting aside normative considerations, the direct involvement of the highest authority at the earliest stage was obviously conducive to the rapid first-step resolution of problems as well as to a high degree of informal and oral activity.

The role of the staff as it operated at Company 4 was in harmony with the philosophy of union-management accommodation as it existed at the company and should not be viewed simply as a procedural arrangement. The direct and informal procedure was an integral part of the character of the union-management relationship. Harmony in the relationship also must not be exaggerated. The record was flawed by the considerable and successful use in the low-rate silicon department of slowdown tactics in the face of a management attempt not to transfer high incentive earnings to new equipment in the new department. But staff authority in grievance resolution was decidedly high at Company 4 and the staff set the tone for the union-management relationship. Low formal grievance rates were the consequence.

The labor relations staff was also involved in the early stages of resolution at Companies 3, 5, and 2 (steel department), but not in the same manner and not to the same degree, except possibly at Company 3. In these companies the labor relations staff operated through a process of consultation. Again, however, the process of consultation in effect at Company 3 produced a higher degree of staff control than that operating in Companies 5 and 2 (steel department). Consultation at Company 3 by line management with staff was typically a "before the fact" type of exchange and came close to being a routine procedure. Consultation at Company 5 was less frequent and was more on a "when requested" basis. Company 2 (steel) was more nearly comparable to Company 5 than to Company 3.

The resolution process began in all of these companies with the employee or his union representative, or both, usually bringing a complaint or problem to the general foreman. If the general foreman were certain of the proper response, he could settle the issue. If he were uncertain, he would consult with the superintendent or assistant superintendent who, if unsure as to the appropriate answer, would consult with the labor relations department representative. If the answer to the informal complaint or grievance was unsatisfactory, and a written grievance was filed, the labor relations department provided the written response for the superintendent.

At Company 3 consultation was a frequent, if not routine, practice and was typically a "before the fact" kind of consultation; that is, no decision was given by line management until after consultation. At Company 3, industrial relations staff made regular daily or twice daily visits to operating departments. Problems tended to be brought to the staff on a "What should I do?" basis. This is not to say that line management played no role, but, compared to Company 4, staff exhibited a high degree of authority and appeared to call the shots in a higher proportion of cases than at Company 5 and at Company 2 (steel). It could be debated whether staff authority was any lower in Company 3 than in Company 4 even though the method of operation was different.

Line management had a higher degree of shared authority in Company 5. In this company labor representatives were available to foremen, general foremen, and at the superintendent level. This company was the only one that was making a concerted effort to involve foremen, as distinct from general foremen, in the resolution process. For example, in the high-rate coating department the foremen as well as the general foremen consulted the labor relations representative responsible for the department. Consultation was much more on a "when needed" than on a routine basis and initial decisions would frequently have been made so that the question put to staff was frequently a "what next?" type of question. At Company 5 line management frequently acted without consultation, though the distinction being drawn between Company 5 and Company 3 is clearly a matter of degree. At Company 5 a labor relations representative sat in on the Step 2 meetings at which the superintendent heard the case and the staff later provided a written answer for the superintendent. At Company 3 the staff did not sit in on the Step 2 hearing but in essence prepared the company position including the superintendent's written response. Line management at Company 2 (steel department) consulted labor relations early and often but not so routinely as at Company 3. Company 2 (steel) was reasonably comparable to Company 5, though the line-staff relationship at Company 2 (both departments) was less conscious and deliberate than at either Company 3 or Company 5.

The labor relations department at Company 1 was generally not involved in the early stages of the problem-resolution process at all. The department became involved only at step 3, where the hearings and the investigation were solely their responsibility. The explanation of their role was twofold. First, an extremely heavy Step 3 workload did not allow labor relations personnel much time for involvement on the lower rungs of the procedure. More basic than this, corporate industrial relations management believed that resolution at Steps 1 and 2 should be a function of line management. At one time, labor relations staff sat in on some Step 2 hearings. This practice was discontinued because labor relations management believed that line management was not doing a satisfactory investigative job and was deferring too many decisions to labor relations staff.

Some individuals in labor relations at Company 1 believed that efforts on their part at the lower steps would reduce the number of appeals to Step 3. This was not idle speculation on their part. Since ultimate management authority resided in the labor relations department, except for wage incentives in which the industrial engineering department had heavy responsibility, appeal to Step 3 was highly probable in the absence of earlier consultation unless line management was unusually strong. At Companies 3, 5, and 2 (steel), where the union knew that labor relations had been consulted early, there was less propensity to appeal grievances to Step 3 to obtain formal consideration by the staff.

The system of line-staff relations at Company 1 also tended to produce more uneven results since it relied heavily on the line. Where the line was strong and conditions were otherwise favorable, as in the coke ovens, no other system would have been superior. Where the line was weak and conditions were otherwise unfavorable (the bricklayer department), lack of staff involvement aggravated managerial weakness.

In the Company 2 wire department, labor relations did not participate until a written grievance had reached the third step of this four-step procedure. One of the reasons for the difference between the steel and wire department procedure was that a labor department position was vacant in the wire department. However, wire mill line management showed a longstanding bias against consultation with the labor relations department that bordered on outright mistrust. The difference in attitude of managements in the wire mills and steel mills was one of the factors contributing to the higher grievance rate in the wire mill. The number of formal grievances in wire as distinct from steel, and the number of appeals to the third step, was in part necessitated by the absence of an informal route to the labor relations department. As discussed in Chapters 2 and 3, there were other reasons for the differences in the relative grievance rates.

Although this study probably does not demonstrate clearly that any particular line-staff relationship is superior, it indicates that some tradeoffs are involved? Generally speaking, the greater the labor relations staff involvement at the earliest stages, the lower the grievance rates and the quicker the resolution of problems, even though staff involvement can result in some short delays if line management is unable to respond to a complaint either because it does not know the answer or does not have the authority to provide an answer. Also early staff involvement encourages informal first-step resolution. On the other hand, where staff does not participate at the early stages, more problems are put into writing and pushed up to the point in the procedure where they receive formal staff consideration. This results in a higher rate and slower as well as more formal resolution.

* Although there are obvious advantages to early staff involvement, the closer the procedure comes to domination by the staff the more probable are certain disadvantages which this procedure entails. Line management may begin to regard the resolution process as beyond the bounds of their job and pass the buck entirely to the staff. If this happens, management may lose a substantial element in the control of efficiency and operating conditions. The balance in shared authority will be further considered in the next chapter. What is clear at this stage is that a considerable range can exist and that management should give conscious attention to achieving an appropriate balance of authority both in general and in relation to departmental variations in environmental and union conditions.

Two other aspects of management organization need to be considered. One involves the structure of the line management organization as it relates to the nature of the task environment. The second aspect centers on the roles of various levels of the line management organization in the resolution process.

In Chapter 2 the influence of environment on the grievance rate and the resolution process was discussed. Important differential effects in the grievance rates and in the resolution process were associated with differences in task requirements and conditions. The elements in the nature of employee tasks and work environment which led to instability also generally meant that the managerial task was less difficult in more stable and pressure-free environments. Not only did certain environments create more employee relations problems, but the entire management task—planning, scheduling, organizing, motivating, controlling—was much more difficult.

Chapter 2 discussed in detail how differences in task environments influenced the grievance rate and the resolution process, so no attempt will be made here to review how these differences could make management's grievance resolution task more difficult. However, it can be noted how task requirements differentiated the management job in other ways.

In some departments quality and quantity of output were more crucial than in others, sometimes simply because they were more easily measured. In these departments quality and quantity were operator-controlled and a key supervisory responsibility. By contrast, in the low-grievance coke ovens department at Company 1 and the utilities department at Company 5, there was little concern with quality or quantity of the product. Consequently in these departments there was more time to devote to other activities and fewer pressures on the work force for quantity and quality. Both low quantity and quality pressures and greater availability of supervisory time contributed to a lower grievance rate.

Maintenance departments, such as those studied at Companies 1 and 4, had quantity and quality pressures, but the speed with which maintenance work was completed and its thoroughness were difficult to measure, so management in these departments may have faced fewer organizational pressures over quantity and quality. However, the ambiguity of the maintenance task with regard to quantity and quality may have made the management job more difficult in some ways. For example, how can a maintenance manager know whether or not a repair job was completed with dispatch? In the absence of standards, which are difficult to apply to maintenance work, much depends upon managerial judgment and experience.

In general, in steel, product quality is more visible as the product moves closer to the finished state. Thus, pressures are greater on management for quality in finishing operations than in basic manufacturing. At Company 2, where both steelmaking and steel-fabricating operations were studied, maintaining quality standards was a much greater part of management's job in the high-grievance-rate fabricating departments. Also, it should be noted here that as goods move closer to the finished state, the problems and consequences of waste of, or damage to, the product become more acute and costly. These add further to line management's responsibilities in some departments.

Differences in the quantity and quality of the managerial task are related to differences in the variability of the task, which also affects the nature of the managerial job. Basic manufacturing operations in steel—blast furnaces, coke ovens, open hearths, BOFs—are chiefly repetitive. Some intermediate operations: slab mills, blooming mills also tend to be repetitive. Input is relatively homogeneous. There are no changeover points at which quality can be affected. Scheduling, manning, and controlling of the operation are relatively easy. The low-grievance-rate departments at Company 1 (coke ovens), Company 2 (steel mills), and Company 5 (utilities) all fit into this category.

Finishing operations in steel are less repetitive than basic steel manufacture. For example, in producing steel sheet and strip there are a variety of dimensions in the finished product—gauge, size (width and length), surface finish—which all require adjustments during the manufacturing operation. The most variable task of all, however, is maintenance work. Aside from routine or preventive work, maintenance and repair seldom face the same task twice. Management in maintenance operations faces problems in scheduling, coordinating, work assignment, planning, and control that management in production operations does not confront. Maintenance management frequently has to respond to crisis situations and it must coordinate the work of different craft groups. A maintenance group in the high-grievance-rate wire mill at Company 2 provides a good illustration. A local work rule required that seniority determined job assignments. A worker could refuse to do a job and insist that it be given to a man with less seniority. The foreman not only had to develop a schedule on the basis of who was capable of doing the work, but also on the basis of who was likely to accept the assignment. Workers frequently would refuse assignments, and the whole work group would sit around until the work was reassigned.

Not only is maintenance work extremely variable, but the locus of the work is changeable. Maintenance management in the study in the steel industry generally could not directly supervise each job constantly. Supervision was generally always on the move just to keep tabs on individual maintenance operations. Maintenance management simply had to spend time getting from one place to another, thereby losing time that should have been devoted to the basic management tasks. The difficulty of the management task in maintenance was certainly a significant factor in the higher grievance rates that were usually found in maintenance operations in the steel companies studied.

In one of the locations investigated, the management organization in the maintenance operation and the locus of decision making compounded the problems posed by differences in the nature of the work. Generally speaking, each of the units studied was headed by a superintendent, except for the bricklayer department at Company 1. The entire mechanical maintenance operation at Company 1 was headed by a superintendent, and the bricklayer department by a general foreman. Thus, roughly the same number of employees were supervised in the bricklayer department as in the coke ovens department, but by a lower level of management. More important in this instance, the relative managerial ability and authority of the individuals matched their organizational position. The management task was more difficult in the bricklayer department than in the coke ovens department, but the management structure and managerial ability did not reflect this fact, and actually compounded the difference. This situation resulted from applying a standardized organizational structure to dissimilar task objectives. It should also be noted that top mechanical maintenance management, the level of management above the bricklayer general foreman, corresponded organizationally to that of top coke ovens department management, and was widely regarded in this plant as indecisive and not on a par with coke ovens management. This is a separate variable, however.

The other maintenance department studied, that at Company 4, was organized differently from the one at Company 1. It was headed by a manager, and there were superintendents under him who were responsible for approximately the same number of men as their counterparts (that is, superintendents) in production operations. Although the managerial tasks in the maintenance department at Company 4 were more difficult than in production departments, the management organization of the maintenance department was at least not inferior to that of production. However, organization still did not adequately reflect differences in task requirements.

In only two of the departments studied did the foreman have any meaningful role in the problem-resolution process. This fact was surprising since contractually at least the foreman, as the lowest level of management, filled the initial management role in the resolution process. In practice, however, the next management level, the general foreman, filled the postulated role of the foreman in the resolution process. If any employee had a complaint, he generally took it to the general foreman, with or without his union representative.

Only in the coke ovens department at Company 1 and in the coating department at Company 5 (and not to the same extent in Company 5 as at Company 1) were the foremen involved in the resolution process. Similarly, only in these two departments did the foremen participate in disciplinary action against employees. Foremen here could adjust schedules, approve or arrange shift changes, and handle overtime problems. In other departments these activities were dealt with at higher levels of management. In Company 5 the expanded foreman role appeared to reflect corporate policy and practice.

In some departments, notably in the steel mills at Company 2, in the structural mills at Company 3, and in the silicon department at Company 4, the general foremen were available in the work place to perform these tasks and to handle these problems. If there was not a general foreman at all times on all shifts, that is, if they only worked on the day shift, their hours tended to overlap the end of the night shift and the beginning of the afternoon shift so that they were available to employees who worked on shifts other than the day shift.

In the two maintenance departments studied, those at Company 1 and Company 5, and in the Company 2 wire mill as well, all of which were high-grievance-rate departments, the general foremen were not always available in the work place, thereby blocking the speedy resolution of problems. In the maintenance departments the dispersed nature of the work meant that the employees were frequently physically distant from the management representatives with the authority to solve problems. In some cases not even a foreman was present in the work place, but even if one were available, he would not have the authority to handle questions or problems. In the departments studied in the wire mill at Company 2, the general foremen were responsible for more than one operation and as a result were not in one particular location all the time. Here the same problem of physical distance separated the general foreman and the work force as in the maintenance departments and the same lack of authority prevented the foremen who were on the scene from handling problems.

The utilities department at Company 5 was a special situation. The work there was not always performed in a variety of places. That is, the locus of the work did not change, offering a contrast to maintenance operations. Supervisors with problem-solving authority were in at least some of the locations at all times, but only on the day shift, so that problems had to be taken care of at that time. This meant that employees on the other shifts handled their own reporting on and reporting off as well as their own distribution of overtime. Consequently, employees in this department made decisions that in the other departments studied were made by management. In the other departments, decisions made by management were challenged on occasion, both informally and through the grievance procedure. Since these decisions or acts were made by utilities department employees themselves, they were not challenged.

Although the absence of supervision on two of the three shifts in the utilities department at Company 5 was unique in the steel companies studied, in a number of other departments observed supervision appeared to be "lighter" on the second and third shifts. Sometimes this was true only in the sense that top department management was not present during these shifts. Grievance rates were generally lower on these second and third shifts, probably for the same reasons as at the Company 5 department, but also perhaps because rules and procedures were less rigidly enforced.

ˑThis phenomenon at least raises the question for management of whether providing increased opportunity for employee initiative or less rigid application of rules and procedures can lead to a reduction in the formal grievance rate. Perhaps it might lead to a re-evaluation of rules and procedures. In any number of situations the strict adherence to rules by employees results in a slowdown. Certainly where the potential for a work-to-rule slowdown exists, rules and procedures might well be re-evaluated to remove potential irritants to union-management or employer-employee relationships.

*The differences noted among departments in the availability of supervision with authority to engage in problem-solving were striking. Unavailability of appropriate supervisory personnel encouraged or necessitated use of the formal grievance machinery. Since foremen were generally not involved in the resolution procedure, and since these low-level supervisors were generally more available in the work place than the higher levels of management who were responsible for grievance resolution, management should probably reconsider the role of the foreman in the resolution process.

Management organization is obviously important in the grievance resolution process. There are several dimensions to this organizational question. First is the question of balance in sharing line-staff authority. If low-level resolution within the operating context is taken as a desirable management objective, then staff advice must be sufficiently available to make unnecessary a formal appeal to Step 3 to reach the locus of staff authority. Adequate lower level staff advice seems essential. At the same time, virtual domination by staff logically has clear dangers to effective management, though these dangers cannot be demonstrated clearly by this study. A balance in shared authority approximating the situation at Company 5 has strong attractions.

Second, if a balance in shared authority approximating that existing at Company 5 is assumed to be appropriate, the substantial differences in the line management task in different departments arising from both environmental and union variables must be recognized before implementation. In some difficult-to-manage departments the application of rather standard organizational structure leaves line management too thin to perform its role effectively in the resolution process. Strengthening the role of foremen is a partial answer, but the typical line structure appears to give inadequate recognition to the inherent difficulties in managing some functions and departments. Either the standard line organization must be modified or special staff support provided in difficult-to-manage departments. Certainly organizational structure should not be allowed to handicap low-level resolution and some adjustments in particular departments in both line and staff structure may well be appropriate.

Management Leadership

Different types of leadership behavior are difficult to deal with analytically. To some extent different leadership styles appear to be an essentially independent variable reflecting differences in personality. But reactions to other variables can also be a factor. In the union case there is evidence of a high degree of independence in the union leadership variable even though these differences have not been probed in any depth. Militant and aggressive union leaders (radicals, politicians, and advocates) had special motivations that created high grievance rates. Problem solvers also had their particular personality characteristics. Yet to some degree union leadership behavioral differences are reactions to variations in the physical environment and in managerial conduct.

Three types of management leadership can be distinguished: *inactive, problem solver,* and *autocratic.* But to a much higher degree than in the union case, these differences in behavior generally appear to be responsive. Although personality differences are found within management, the range of difference appears narrower than in the union case.

Inactive management—that is, inactive so far as relations with the union are concerned—appeared particularly to be responsive behavior. Almost all foremen fit this classification. Organizational tradition or mandate limits the scope of foreman activity. Exceptions occurred in Company 5, particularly in the coating department, which reflected an active policy to encourage foreman participation, and, in the Company 1 coke ovens, a relatively unusual situation of strong line management coupled with fairly inactive union leadership. Inactive managers at the general foreman and the superintendent level were also found in situations where they were backing away from strong union pressures.

Bricklayer department management (and indeed mechanical maintenance department management) at Company 1 was generally inactive. They went through the motions of problem resolution, of meeting with the union, but did not really make decisions or really attempt to resolve important questions. They preferred to pass the buck to labor relations, which in turn had a hands off policy toward the work place. Therefore management was free to handle problems as they chose, but in this case it also meant that management could back away from problems and problem resolution.

In an extreme case (and this happened occasionally at Company 1), weak management personnel would make decisions and then reverse themselves under union pressure. In the Company 1 coke ovens department, some time before this study, management did not discipline at all and did not maintain work standards. After top management discovered what was going on, a totally new industrial relations climate was created by new departmental management.

The *problem solver* is the second category of management leadership. Although staff management was heavily involved in both Company 3 structural mills, in the Company 2 steel mills department, and in both departments studied at Company 5, line management also was typically concerned with the problem-solving process. In these places, line management was responsive to informal complaints and challenges, and, leaning heavily on staff advice, worked to settle issues informally. These managers behaved as problem solvers. The managerial problem solver was just like the union problem solver. He was a careful listener who attempted to differentiate valid complaints from invalid ones. Since management decisions or actions were often the basis of employee complaints, management problem solvers were willing to admit their errors or mistakes. Some line managements actively consulted with the union representatives prior to making final decisions and in this manner headed off challenges by achieving tacit consent to decisions before they were actually made or promulgated.

In the steel mills at Company 2 and in the coating department at Company 5, even though staff advice was available and used fairly frequently, line management exhibited a strong desire for resolution at the department level without labor relations department involvement. Line management believed that the individuals at that level were in the best position to make the decisions required. The managements in these locations could be called more active problem solvers than in the others mentioned. The management in the coke ovens department at Company 1 was also a very active problem-solving group and typically acted without labor relations staff participation. Lower line management representatives were more active problem solvers in these three situations than in any other departments studied.

At times the desire of line management to settle problems informally was the result of what was almost a fear of written grievances. These individuals tended to view the emergence of a written grievance as some sort of personal failure. In one instance in the open hearth operation of the Company 2 steel mills department, this phobia on the part of the supervisor was used by the union grievance committeemen to their advantage.

• The third type of management leadership found could perhaps best be called *autocratic.* Autocratic managers tended to believe in their exclusive right to make decisions and treated any union or employee challenge as incorrect, irresponsible, or even illegitimate. Even informal challenges generally received a negative response, which of course meant that challenges tended to be formalized as grievances, which in turn were not responded to positively at the lower steps of the grievance procedure. The attempts of labor relations at modifications of decisions in the higher steps were resisted or at least disapproved of by these managers. Examples of this type of leadership were found in the Company 2 wire mill and in the Company 4 maintenance department. Though managers in these departments did not completely fit the description of an autocrat, their behavior tended to move in this direction, in contrast to a real problem-solving type of behavior, and was likely to provoke formal challenge through written grievances.

In every location studied, autocratic tendencies were modified by contractual and labor relations staff pressures. According to the old-timers who were interviewed in the course of this study, the autocrat was the standard type of management leadership in pre-union days, though no doubt this stereotype is carried somewhat beyond reality.

Although it might have been useful, a far more complete analysis of management (and union) leadership styles than this simple behavioral classification was not attempted because of the complexity of innumerable influences. There are various multiple-level policy and organizational influences impinging upon each individual acting in a particular technical and union environment. More sophisticated labels do not appear to be warranted under these circumstances.

Styles of management were influenced in part by labor relations staff involvement. To put it another way, labor relations staff activities, under a company policy that recognized union legitimacy and provided a role for the union in decision and rule making, placed constraints on the range of line management behavior. But the role of staff was not totally consistent in each plant and some management leaders were more independent of staff than others. Also, task requirements and conditions and the structure of line organization implicitly created different staff responsibilities. Finally, policy constraints and differences influenced the challenge rate and the resolution process.

The marked contrast between the relative importance of management leadership differences and union leadership differences on the challenge rate and the resolution process should be remembered. In the union case the locus of departmental decision-making power tended to be concentrated in one man, whose leadership style was thus of crucial importance. Within management, decision-making power was much more diffused. If a given management leader was inactive, other line or staff individuals assumed leadership. Consultation within line and between line and staff diminished the dominance of single individuals, as did the more pronounced policy-making role of management as contrasted with the union. Despite these statements, low-grievance-rate departments tended to be under the influence of policy-oriented line management personnel who tended to be problem solvers. High-rate departments had essentially inactive line leadership thus increasing staff responsibility, somewhat autocratic leadership, or strong problem solvers as in the Company 5 coating department. Although the influence of strong problem-solving and policy-oriented management leadership, such as was found in the Company 1 coke ovens, does not stand out so dramatically as does aggressive union leadership, not all management leadership should be interpreted as responsive behavior. Unmeasured foreman and other line management leadership behavior strongly influences employee morale and indirectly grievance-related behavior.

Management Policies

• Management responds to pressures or anticipated pressures from the environment and from the union. But some management responses were more policy oriented than others. Policy can be described as a predetermined formal and informal framework for making decisions. The lack of a policy-oriented response on the part of management usually means inconsistency, which in turn invites union challenge—challenge that can often be successful. Also, the role of labor relations staff in the resolution process is a substantive policy decision. The decision to allow line management to handle problems on their own, at least at the lower levels of the resolution procedure, heightens the importance of implementing policies that will avoid inconsistency and subsequent challenge.

One of the major differences between high- and low-grievance-rate departments was in management's policy on consultation with the union. This policy, since it strongly relates to both the general relationship with the union and management's perception of its own role as a decision maker, is a basic substantive policy. Quite often it was not really viewed in this way by management individuals, who tended to call it a "technique." In reality, consultation actively endorsed at the corporate level represents a fundamental policy decision. Both Companies 4 and 5 appeared to have meaningful corporate policies encouraging consultation. However, the willingness of management individuals to consult before taking action was much less a question of corporate policy than it was a function of individual leadership style and this course was most likely to be followed by the active problem-solver type of manager. Kuhn says that: "the guarantee of *ex post facto* justice—impartial and equitable though it may be—is not all that workers or union leaders demand. In addition, they seek the kind of justice whereby managers consult with them before they act." [1]

The present study indicates that consultation did tend

---

[1] James W. Kuhn, *Bargaining in Grievance Settlement* (New York: Columbia University Press, 1961), p. 24.

to reduce the formal grievance rate since it was conducive to informal resolution. When the union was consulted beforehand, the likelihood that union leaders would instigate or support a challenge to the decision decreased markedly. The "problem" was resolved at the decision-making stage, before it really became a problem.¹

At Company 2 (steel department), the policy of consulting the union committee about scheduling led to the elimination of many grievances on this subject. On several occasions, committeemen who had seen and approved the schedule before it was posted refused to support or process complaints about the schedule after it was posted. In the rod mill the union committee was actively involved in planning the changeover to larger sized billets which would be produced by the new continuous caster. In the high-grievance-rate wire mill, supervision did not try to consult with the union about scheduling and was very secretive about changes in the manufacturing process.

At Company 3 the difference in consultation between the Plant A structural mill and that in Plant B was more one of degree than of kind. At Plant B, the union committeemen were consulted more frequently about schedules, overtime, subcontracting possibilities, manning, and the like, than were their counterparts in Plant A. Such consultations in Plant B did not mean that management would necessarily accept the union's point of view. Frequently consultation meant outlining an anticipated course of action and then inquiring if the union had an alternative plan, and, if not, had any objection to the proposed action.

At Company 5, the management of the low-grievance-rate utilities department tended to consult and to notify the union grievance committeeman to a much greater extent than their counterparts in the high-grievance-rate coating department. For example, a temporary job combination in the utilities department, due to reduced operations, was arranged with and through the grievance committeeman as an alternative to the complete shutdown of an operation. Also, problems of scheduling were generally discussed with the committeeman before decisions were reached. Management's philosophy in the coating department was that a manager should make the decision he thought was best under the circumstances, without reference to any other opinion. That philosophy extended to staff groups as well as to the union. Coating department management was an aggressive group, more likely to think in terms of "management's rights" than were their counterparts in the utilities department.

At the Company 1 coke ovens department the union was generally inactive, and it did not frequently object to any management decision. This was the only department where the union appeared inactive but it was also the department in which management took the most pains to observe the agreement carefully. In the bricklayer department, where the union was active and aggressive, management did not follow a policy of consultation. In the bricklayer department, as well as in the other high-rate departments, the attitude of the union representatives made consultation difficult and at times unpleasant for management. The lack of consultation by management was thus obviously not independent of the character of union leadership. Nevertheless, management in some departments and companies was more committed to a policy of consultation than in others.

At Company 4, the personality of the union committeeman in the maintenance department acted to reduce the amount of consultation in the silicon department. Top management in the maintenance department was young and aggressive. Top management in the silicon department was of a similar bent and could perhaps be said to have been in the process of modifying their policy on consultation as a result of experience. In an incident which occurred at the time of the study, the failure of line management to consult with the employee involved or the union on a temporary move-up on a slitter resulted in a written grievance which labor relations believed could have been avoided had line management followed a policy of consultation. For acting without consultation and for breaking a recently negotiated local working rule in their actions on the temporary move-up, line management was criticized by labor relations management and urged to consult more in the future.

|Consultation can be looked at from two points of view: first, as an important part of corporate policy and, second, as an adjunct to an active problem-solving leadership style.| Company 4 and Company 5 appeared to have more active corporate policies on consultation than the others, though our study is not broad enough to confirm this with certainty. What is much clearer is the use of consultation by problem-solving management leadership. A reinforcing relationship between problem-solving management and union leadership tended to be created by the use of consultation. Yet little attempt was made at consultation—and it probably would not have been possible—in the high-rate departments. Where practiced, consultation appeared not to imply any loss of management's ability to manage, and it clearly reduced the grievance rate and increased the informal resolution of problems.

Of all the operating policies, those on incentives are among the most important and most difficult for management. The basic policy decision is, of course, whether to use an incentive system on a particular operation. In steel, management's attitude in this regard has been somewhat restricted by the 1969-70 industrywide arbitration decision and subsequent company-by-company agreements on incentive coverage. Once this basic decision on use of incentives is made, management is then faced with the problem of worker pressure on incentive payout—coercion which frequently comes in the form of slowdowns. Some companies have taken strong stands on production standards and have successfully weathered the pressure. Others took strong stands but were later forced to retreat or capitulate.

Management action in the silicon department at Company 4 was such a retreat. This new department was a consolidation of silicon operations located elsewhere in the plant and at the same time involved the installation of new equipment. Employees from the old silicon operations were transferred to the new ones as they went on line—a process which was still under way at the time of the study. At the time operations were begun, the silicon department's policy on incentives was that incentive earning in the new department would be limited to approximately 135 percent of standard hourly earnings and would not be set at the out-of-line level of 170 percent to 190 percent of standard which had developed over time in the old operations.

Under pressure, in the form of slowdowns, which occurred on every piece of equipment installed, management backed away from its initial policy and earnings of the employees working on the new equipment were almost as high as they had been on the old. Of course, the new equipment required fewer employees, so management, even with an incentive retreat, was still receiving cost benefits from the new equipment. Correcting a high-yield incentive is obviously difficult.

The labor agreement at Company 4 provided for the arbitration of incentive questions. The incentive on a tandem mill for rolling silicon steel was arbitrated after a slowdown which lasted for more than a year. The arbitrator adjusted some elements in the standards upward, but the employees on the tandem mill were still unsatisfied as the study ended, since their earnings were generally lower than other units in the department.

One of the slowdowns in the silicon department, in its shipping operation, had at least the tacit approval of line management. The location of the department had shifted as new units were placed in operation in the department, and the physical relationship of different parts of the shipping department to one another had changed. Incentive earnings had declined as a result, but department management could not get the industrial engineering department to agree to a restudy of the shipping operation. A slowdown by employees resulted in a restudy. The slowdown continued until adjustments were made; but the measures went beyond those line management believed necessary. The real source of the discontent among shipping employees over their incentive was the relatively high levels of earnings elsewhere in the silicon department. These high earnings in turn resulted from the use of pressure tactics and were related to earnings on the old silicon finishing equipment. Thus, problems of incentives in the silicon department were related to previous developments and management response over those years.

At Plant A of Company 3, a slowdown had been in progress in the shipping department of the structural mill for over six months at the time the study was made there. The situation was quite similar to that in the silicon department at Company 4, although it was compounded at Company 3 by the attitude of union leadership. The shipping department incentive plan in the Plant A structural mill was based on the number of items shipped. Over time, the size and weight of the items shipped had increased; steel was shipped in larger sizes and the number of items shipped was lower. As a result, incentive earnings had declined. Management sympathized with the shipping employees and had offered a revised incentive plan. The union leadership had refused the offer, but failed to report to the employees involved that an offer had been made, and the slowdown was unresolved as the study at Plant A was concluded.

The dissatisfaction of the shipping employees stemmed, just as it had at Company 4, from their relative position in incentive earnings. Earnings on some structural mill units were substantially above standard hourly wages, while in others, including shipping, performance was below standard. Incentive yields in the Plant A mill had gone above the level advocated by industrial engineering because of a willingness on the part of line management over the years to yield to group pressure on incentives.

Management in the Plant B structural mill had not followed this policy and, as a result, the top incentive yield there was only modestly above normal. Incentive earnings were much more closely clustered about the desired 135 percent level than they were in the Plant B structural mill.

The Plant B mill had experienced wage-related slowdowns, but they had been more easily broken than those in the Plant A mill. Unlike the Company 4 silicon department, where almost all incentive problems involved or resulted in the use of pressure tactics and no grievances, the two Company 3 structural mills had incentive-related grievances. As might be expected, incentive and other wage-related grievances were greater in number in the

Plant A mill than in the Plant B facility, and were a larger proportion of the total number of grievances there than they were in Plant B. The rate of incentive grievances per hundred employees per year in the two mills was as follows for the years 1964-1967:

	1964	1965	1966	1967
Plant A structural mill	1.3	0.9	0.9	0.7
Plant B structural mill	0.9	0.5	0.5	0.2

The rate in the Plant B structural mill was roughly half that of Plant A. In Plant B incentive grievances were a smaller proportion of the total number of grievances than in Plant A. Incentive grievances as a percent of total grievances in the two structural mills for the years 1964-1967 were as follows:

	1964	1965	1966	1967
Plant A structural mill	56.1	17.6	10.5	4.3
Plant B structural mill	10.5	10.5	5.0	3.5

The incentive problems at Company 5 and Company 1 were discussed in detail in Chapter 2. It should be noted that aggressive management action stopped the one slowdown in the Company 5 coating department. The incentive yields there had not been permitted to get out of line with the desired level of 135 percent of standard hourly earnings, but line management had also acted to insure earnings at that level. Incidents of managerial pressure on industrial engineering to change standards to guarantee this level of earnings were seen, as well as suggestions from line management to the union that grievances be filed to insure retroactivity on incentive changes.

At Company 1, the management policy of having incentives for maintenance employees insured in some case relatively good wages for maintenance employees as compared with production employees. However, the performance of most maintenance groups was below standard, and the incentive system was extremely difficult to manage, as was indicated in the last chapter. As a result, company policy at the time of the study was not to expand the use of maintenance incentives at the plant studied or in other Company 1 operations.

Also of interest is the fact that line management in the bricklayer department had no role in the administration of the incentive system there. Rates were established by a rate setter from the industrial engineering department who was permanently attached to the department. Exceptions and special allowances were made by the rate setter, and union representatives held discussions about incentives with him and not with line management. Of course at every location where incentives were in effect, line management shared authority for changes with the industrial engineering staff. Line management could exert pressure on industrial engineering to restudy jobs and to make adjustments, but the actual changes in the incentive standards were made by industrial engineering, as line management generally lacked the expertise to administer an incentive system. At times, though, changes in standards were made over the strenuous objections of industrial engineering, as was the case, for example, in the Plant A structural mill at Company 3. Industrial engineering placed the blame for the distorted nature of the incentive system there solely on line management.

None of the companies studied was free from incentive problems and only two plants did not have rather severe inequities in incentive earnings. But the high-grievance-rate departments tended to have more severe incentive problems than the low-grievance-rate departments primarily because environmental differences made incentives more difficult to administer in those departments. Also, incentives tend to reinforce conflict and high grievance rates. Incentive inequities are an open invitation to aggressive union leadership and encourage the use of slowdowns. Union pressure tends to create additional injustices. Negotiating standards and yielding to pressure create a vicious circle in which some group is always lagging. However, one high-rate department, the coating department in Company 5, maintained good incentive administration and did not yield to pressure. Incentives tend to cause and to reinforce high grievance rates just as consultation reinforces low rates.

Since discipline policies are also substantive, they have an important effect on the grievance rate. The use of systematic discipline systems offers the best evidence that consistency in management action, that is, policy-oriented management action, can reduce union challenge. The two departments studied at Company 1 presented the best contrast between disciplinary policies. In the coke ovens department, progressive discipline systems were in effect for absenteeism, poor workmanship, safety, and equipment abuse. The existing policies were the result of a serious situation that had developed in the coke ovens department prior to the advent of strong supervision and the creation of policies. Under the progressive discipline system in use at the time of the study, records were carefully maintained and discipline was generally automatic and fairly frequent. The foremen played a major role in the administration of the system, since they represented management in all contacts with employees about discipline. Ninety percent of the written grievances in this low-grievance-rate department were over discipline. Because of the progressive nature of the discipline, few of these grievances went to arbitration, and the few compromised by management were whether to give a deserving employee a second chance, or, according to coke ovens department management, were traded to the union by the labor relations department in return for the withdrawal of another grievance. Not all of the disciplinary actions taken by management were grieved, in part because the progressive nature of the programs had built-in warnings of increasingly severe penalties. The union could only argue over facts and not over degree of penalty. Indeed, the union in this case became almost an adjunct of management in working to get employees to improve their attendance and workmanship so that their eligibility for discipline would be reduced.

Although the need for a progressive discipline system appeared as great in the bricklayer department as in the coke ovens department, one did not exist there. As a result, a higher proportion of disciplinary actions were contested —a larger number successfully—in the bricklayer department. As a result of having disciplinary penalties reversed, line management was somewhat demoralized and reluctant to discipline. Discipline was a source of conflict between line management and the labor relations department, with the former complaining that labor relations would not uphold their penalties and labor relations in turn saying that they could not do so because line management was not consistent. Both management groups were probably right in part. Line management was not especially consistent, but even where they had made serious attempts, particularly in connection with a difficult union leader, management had reversed some of the penalties for the purpose of improving or fostering good labor relations.

At Company 4, the same line-staff problem of discipline

was found. The number of disciplinary penalties meted out by management was much smaller in Company 4 than it was in the departments studied at Company 1, and thus the number of union challenges was smaller. Only in the maintenance department were written grievances filed protesting disciplinary penalties, and all of them were granted. This irritated line management, although the reasons for the reversal were either inconsistency or mistakes on their part and were not the result of any particular decisions by the labor relations department.

In other plants studied, disciplinary policies did not differ so radically between high- and low-grievance-rate departments. At Company 5 the policies were basically the same, but because of differences in the nature of the work and in the work force, discipline tended to be used more in the high-grievance-rate coating department than in the utilities department. Production employees in the coating department were held responsible for the quality of the product, and disciplinary penalties were used when failure to meet quality standards was the fault of operator error. The utilities department had no quality concerns, and thus there was little need to discipline. Then, too, the nature of the work in the utilities department required less effort and a higher caliber of employee, which resulted in fewer absenteeism problems than in the coating department.

The greater number of union challenges to disciplinary penalties imposed in the coating department were in large part responsible for the differences in the grievance rate between that department and the utilities department. There were more penalties in the coating department, and the union probably challenged a greater percentage there than they did in the utilities department. Thus, the real differences in the grievance rate in the departments studied at

Company 5 can be attributed more to differences in the environment and in union leadership than to differences in policy. However, it should be noted that coating department management was far more aggressive than were their utilities department counterparts. More assertive management might well have imposed more disciplinary penalties and had more grievances.

Both structural mill departments studied at Company 3 used progressive discipline systems. However, owing to the environmentally influenced differences in the work force (mentioned in Chapter 2), management in the Plant A mill found it necessary to discipline employees more frequently than did their opposite numbers in the Plant B mill. The rate of union challenge was about the same in both mills, however, and all disciplinary actions were not challenged.

In Plant A, grievances over initial discharge actions were almost always granted, particularly those dismissals resulting from a culmination of a series of disciplinary actions against an individual. The second discharge was almost always final, and although the union might grieve the second action, it did so only as a matter of form, since the issue rarely, if ever, went to arbitration. The first discharge (or more properly, suspension with intent to discharge) acted as a warning to both the employee and the union that the next violation would result in dismissal. The union committeeman and the employee's fellow workers generally tried in earnest to help the suspended employee correct his behavior so that he would not be subject to a second discharge.

Progressive discipline systems tended to break down somewhat in high-rate departments. This was particularly true of the bricklayer department in Company 1. It was not true of the coating department in Company 5. Again, as with incentives, discipline was difficult to administer in the high-rate departments because of their operating characteristics and could serve to reinforce the focus of conflict. Loss of managerial control was evidenced by ineffective administration in both incentives and discipline, albeit typically in the face of aggressive union challenge.

Summary

Three aspects of management influences on problems, challenge, and resolution have been considered: management organization, management leadership, and selected management policies. All three variables seemed distinctly more interdependent than was the case in the union analysis; in addition, they appeared quite responsive to and interrelated with environmental and union variables. Policy and organizational differences were corporate rather than departmental in character—so far as deliberate differences were concerned. There were, however, unintended and unplanned departmental variations as well as variations that resulted from deliberate policy differences. These interrelationships will be an important consideration in the concluding chapter.

Although three line-management leadership styles were observed—inactive, problem solving, and autocratic—differences in leadership styles were much less marked and distinctive than was true in the union analysis. In part, this was a reflection of shared and diffused decision-making authority. Individuals stood out less clearly than in the union case and also were subject to organizational and policy constraints in making decisions and taking action. Also, in considerable part, management behavior was responsive to environmental, union, and organizational constraints and pressures, rather than being a clear reflection of personality differences.

Almost all foremen were inactive because of their limited organizational role. There were two departmental exceptions in which foremen had meaningful roles. The coating department in Company 5 appeared to reflect conscious corporate policy; the second exception reflected strong line management leadership in a stable low-pressure environment with inactive union leadership. Some line managers, general foremen, and superintendents were inactive, though they may have gone through the procedural formalities to escape or avoid problems or because of a very complex managerial task. The most notable circumstance was avoiding a difficult union situation by passing the buck to labor relations. Inactive leaders tended to create a higher level of formal grievance activity and to necessitate staff resolution at Step 3 and above.

Most line management leaders were problem solvers, some more actively so than others. In part this was a response to expected corporate behavior; in part it was a response to problem-solving union leadership and favorable environmental conditions; and in part it was a personal leadership style. All of the low-rate departments tended to have problem-solving line management leadership, as did two of the five high-rate departments. These two high-rate departments were led by strong line management individuals. Other high-rate departments had autocratic or inactive leaders. Autocratic leaders were strong "management rights" individuals who were also responding to a difficult union or environmental task situation, not by backing away, but by maintaining line management control.

The analysis of management leadership presented here

is more superficial than was the case with union leadership, partially because the problem was more complex. There were meaningful differences, particularly among individuals who have been described as problem solvers. Some of these leaders were more active than others and were a strong positive force for high employee morale. Management leaders were not simply responding to union and environmental conditions but were frequently the dominant leaders in the department to whom employees and union leaders were reacting.

Two significant organizational differences were discussed. The first dealt with corporate differences in line-staff relationships. It was clear that early staff involvement at the bottom rungs of the grievance ladder tended to create lower grievance rates, a higher degree of oral and informal resolution, and a higher proportion of lower-step resolution of grievances. If staff were not involved at the lower steps, formal grievances were filed, thus raising the rate; in addition, oral resolution was less frequent, and a higher proportion of grievances were resolved only at Step 3 and above. However, there were subtle differences in the manner in which staff influence and control were brought to bear at the lower steps. This entire subject will be discussed further in the next chapter. There is no clear-cut "best" system for sharing authority.

The second organizational matter of considerable importance was the seeming failure of the organizational structure to fit the line management task in a particular department. Some departments and functions definitely appeared to be more difficult to manage than others, but a more or less standardized line organizational structure was imposed on all departments. Either line organizational structures should be better adapted to the difficulty of the managerial task or special staff assistance should be provided for the difficult-to-manage departments. In this adaptation foremen might well be given an expanded role.

Only selected policies were discussed. Policies for discussion were chosen because they were particularly meaningful for grievance activity and resolution. The widespread use of wage incentives made incentive administration very important. Although there were exceptions, difficult task environments created complicated incentive administration, stimulating and reinforcing a high grievance rate. The opposite was true in the low-rate departments. Consultation tended to reduce grievance rates and to reinforce otherwise favorable conditions. An active, progressive discipline system indicated a policy-oriented management; while lack of an effective discipline policy pointed to relatively poor managerial control of operations. Policies and their administration were in part responsive to union and environmental variables and in part to deliberate guiding principles actively influencing union behavior. The more important of the various interrelationships and feedback mechanisms among all the variables will be explored in Chapter 5, as will the implications of this study for management.

5 Interrelationships and Their Implications for Management

THE THREE PREVIOUS CHAPTERS have analyzed environmental, union, and management factors. Hopefully, this analysis has helped to illustrate in some detail how each of the factors influences the grievance rate and the resolution process. However, the factors do not operate independently and alone. If the grievance procedure is to be truly understood, if differences in grievance rates and in the resolution process between departments, and to some degree between plants and companies, are to be explained with reasonable adequacy, then the environmental, union, and management variables must be seen as interdependent. The three dimensions of the environment —(1) task organization and work environment, (2) technological change, and (3) socioeconomic conditions—interact with the union and management influences in *their* respective three dimensions—(1) leadership, (2) organization, and (3) policy—just as the management and union

Exhibit 4. SUMMARY OF STEEL STUDY

Factor	Company 1 High (Bricklayer)	Company 1 Low (Coke ovens)	Company 2 High (Wire products)	Company 2 Low (Steel making)	Company 3 High (Structural mill)	Company 3 Low (Structural mill)	Company 4 High (Maintenance)	Company 4 Low (Silicon)	Company 5 High (Coating)	Company 5 Low (Utilities)
Environment										
Tight local labor market	Yes	Yes	Yes	Yes	Yes	No	No	No	No	No
Plant size	Large	Large	Medium	Medium	Large	Medium	Small	Small	Large	Large
Community size	Large	Large	Medium	Medium	Large	Small	Small	Small	Large	Large
Civil rights problems	Yes	No	No	No	No	Some	No	No	No	No
Technological change	Yes	No	No	Pending	Pending	In some years	No	No	No	No
Task environment	Unfavorable	Favorable	Unfavorable	Favorable	Same	Same	Unfavorable	Favorable	Unfavorable	Favorable
Union										
Department leadership	Radical	Inactive	No clear pattern tendency to be advocate	Pattern tendency to be problem solver	Advocate	Problem solver	Advocate	Problem solver	Advocate	Problem solver
Internl. staff influence	Low	High	None	None	Low	High	High	High	Low	High
Policy influences	No	Yes	Yes	No	No	Yes	No	No	No	No
Pressure tactics used	Yes	No	Yes	No	Yes	No	No	Yes	No	No
Management leadership	Inactive	Problem solver	Autocrat	Problem solver	Problem solver	Problem solver	Autocrat	Problem solver	Problem solver	Problem solver
Incentive problems	Yes	No	Yes	No	Yes	No	No	Yes	Yes	No
Union consultation	No	No	No	Yes	Some	Yes	Some	Yes	No	Yes
Systematic discipline	No	Yes	Varied	Varied	Yes	Yes	Yes	Yes	Yes	Yes
Difficult task	No	No	Yes	No	Same	Same	Yes	No	Yes	No
Staff influence/control	Low	Low	Low	High	High	Higher	High	High	Moderate	Moderate

interact, and both interact with the environment. This final Chapter discusses these interrelationships and presents some judgments about management policy and practice.

A Synoptic Review of the Study

Exhibit 4 gives a synoptic summary of the steel study, but of course it drastically simplifies and overgeneralizes. Yet it brings out some fundamental consistencies as well as some chance variables and highlights some questions about the relative independence or interdependence of the variables. A brief discussion of Exhibit 4 serves as an introduction to the next section of this chapter—high and low grievance syndromes and interrelations among the variables. Let us turn first to the environmental variables.

Among the environmental influences, the clearest and most consistent variable distinguishing departments with high and low grievance rates is the *task organization and work environment*. However, it has only been possible to contrast task organizations as favorable or unfavorable. Excluding Company 3, in which the task requirements were identical in the two plants, unfavorable task environments typically had certain distinguishing features. In all four of them, employees as individuals had heavy responsibility for quality and quantity of product, which in turn required continuous and close attention to work. This condition necessitated quite close supervision and relatively frequent use of discipline to maintain quality and quantity. These departments thus had relatively high-pressure task environments. In addition, operating conditions gave rise to frequent *routine* changes in product and process. The organizational tasks were thus unstable as well as high pressure. These environments therefore routinely created more problems, had higher problem rates, and, with frequent product and process changes the problems themselves were more novel in character than in the low-rate departments. An unfavorable task organization created not only more problems but also more unusual ones, since no two maintenance tasks were exactly alike.

The favorable task environments were the opposite of these conditions. Workers were not so directly responsible for product quality and quantity, and operations were much more stable. The utilities department and the wire fabrication department offer rather extreme contrasts, but of course favorable and unfavorable task environments were a matter of degree, and grivance-rate differences were by no means related only to task environments. To jump ahead for illustrative purposes, the bricklayer department, which had a very high grievance rate, would never have stood out so dramatically simply because of its task environment had this not been coupled with the use of a complex incentive plan with rather unstandardized tasks, and, even more important, had the union not been led by a radical civil rights union committeeman who brought enormous numbers of discrimination grievances and had the union not had a strong advocate as the grievance committee chairman.

Among the other environmental variables, nonroutine product and process innovations created by technological change were not consistently associated with high-rate departments. The use of new technology, in this small sample of departments sometimes raised grievance rates in high-grievance-rate departments, as in the bricklayer department, and sometimes in low-grievance-rate departments, as in the silicon department and the Plant B structural mills. In this small sample, technological change was a chance variable.

The third environmental variable, socioeconomic factors, was a conditioning variable affecting both high and low departments in a comparable manner, except for Company 3 in which the high and low departments were in different plants and communities. Although general economic conditions were a favorable element in every situation, department size, plant size, and the nature of the community exerted positive or negative influences on both high and low departments, compounding otherwise favorable or unfavorable variables. Civil rights issues were a markedly severe problem only in one high-rate department, though these issues were a modest problem in one low-rate department. In this study, then, civil rights issues and technological change were quite independent variables.

It was not possible to assess all environmental and conditioning variables in this study because the nature of the research was a limiting factor. The general character of union-management relations and its historical evolution no doubt played an unseen role in plant and corporate differences, as would a similar variable applied departmentally. Only at Company 4, with its very low grievance rates, is it possible to assert with confidence that the overall quality of union-management relations contributed substantially to the low grievance rates characteristic of all departments in the plant.

The union variables—(1) leadership, (2) organization, and (3) policy—again reveal in the case of union leadership a consistent distinguishing characteristic of high and low rates. Inactive or problem-solving union leaders were found in low-rate departments, while advocates, politicians, or radicals were found in high-rate departments. To what extent the more militant types of union leaders were in turn the product of difficult task environments and less policy-oriented management is impossible to say. Nevertheless, it seems clear that union leadership variations were significantly independent. Union organization in all departments except Company 2 acted uniformly to concentrate power in one lower level union official and also typically in one higher level union official. Union organization thus acted in a way that greatly magnified the importance of differences in union leadership styles. Exhibit 3 indicates that neither international union staff influence nor differences in union policies consistently distinguished between high- and low-rate departments.

The use of union pressure tactics was often, but not always, associated with high-rate departments. One low-rate department did not have incentives and a second low-rate department had no problems in this area because of an almost automatic and stable incentive system. Most of the others had some problem with incentive administration but the degree of difficulty varied not only with the task environment, but also with the effectiveness of company policy. Some high-rate and some low-rate departments were troubled by pressure tactics, and some had relatively good incentive administration. Union leadership, company policy, and other conditioning variables appeared to bear upon this complex problem.

Management leadership, organization, and policy proved to be complex variables, far more difficult to analyze than the counterpart union variables. Leadership differences did not stand out dramatically except in one or possibly two high-rate departments where, through either inactive or autocratic behavior, line management tended to pass the buck to staff, forcing more formal and higher level resolution. Clearly some low-rate departments had active and effective problem-solving leadership, frequently using consultation in association with problem-solving union leadership. Incentive difficulties and policy weaknesses acted greatly to magnify grievance difficulties in some high-rate departments. Organizational and policy differences will be dealt with throughout this chapter, for these differences determined the relative effectiveness or ineffectiveness of managerial decision making and control of operations.

Low- and High-Rate Syndromes and Interrelations among the Variables

A high-rate syndrome involves, in various degrees, an unfavorable task environment, aggressive and militant union leadership, and ineffective managerial decision making as indicated by leadership, organizational, and policy deficiencies. Obviously these unfavorable environmental, union, and management characteristics interact through various feedback mechanisms. The following outline presents an exaggerated combination of unfavorable characteristics:

I. Environmental influences
   A. Task organization and work environment
      1. High individual worker responsibility for quality and quantity of product
      2. Close attention to work required
      3. Satisfactory quality standards difficult to achieve, requiring close supervision
      4. Frequent routine product and process changes, creating many nonroutine problems and an unstable task situation

5. A work environment that makes informal problem resolution difficult
   6. An incentive system with many nonstandardized tasks and with many earnings and effort inequities
   7. A difficult-to-manage task environment with various possible additional unfavorable characteristics
  B. Technological change
   1. Frequent nonroutine product and process changes induced by technological change, creating job, employment, and wage-rate insecurity
  C. Socioeconomic conditions
   1. Location in a large city environment with frequent union-management conflict and social unrest
   2. A large plant with impersonal relations and long lines of communication
   3. A plant history of union-management conflict
   4. Departmental civil rights issue
II. Union influences
 A. Union leadership
   1. Militant leadership of the radical, politician, or advocate type at both departmental and third-step levels
 B. Union organization
   1. A union organizational structure placing minimal constraints on individual leadership behavior
 C. Union policies
   1. Some special policies carrying all grievances of certain types at least to Step 3
III. Management influences
 A. Management leadership
   1. Line management of an inactive or advocate type, inducing or requiring formal grievance and resolution at Step 3 or above. Little informal problem resolution
   2. Inactive foremen

B. Management organization
   1. Labor relations staff not active below Step 3, again necessitating formal grievances and appeal to Step 3 to reach the locus of staff decision-making power
   2. Labor relations staff swamped by large numbers of formal grievances
C. Management policy
   1. An incentive system with many inequities in earnings and effort aggravated by pressure tactics
   2. A weak and ineffective system of discipline
   3. Substantially no consultation with the union in advance of management action
   4. Other inconsistencies in management action, indicating an inadequate policy framework

Although this outline is exaggerated, it implies various unfortunate consequences for grievance activity and grievance resolution. An unfavorable task environment coupled with ongoing technological change creates a high problem rate. Aggressive union leadership continuously challenges both management inaction and action in response to the high problem rate as well as inconsistencies in managerial decisions. Inconsistent management decisions reflect an inadequate policy framework. The task environment, weak line management, and no staff involvement below Step 3 convert almost all challenges into formal grievances, which are only resolved at Step 3 and above. Ineffective management decision making increases union militancy, which in turn creates further breakdown in the development and application of consistent management policies. An atmosphere of conflict gives rise to emotional rather than rational union-management interaction. The emotional atmosphere is aggravated by civil rights problems, a history of discord, and other conditioning elements. The entire grievance procedure becomes a formal semilegal resolution process centering on rights and obligations, with a low degree of true resolution of employee problems. If some of the above statements were omitted, toned down, or muted, the high-rate syndrome portrayed would approximate the bricklayer department at Company 1.

A high-rate syndrome, however, is a matter of degree, and in fact many combinations of the variables existed in the high-rate departments. The task environment could be more or less unfavorable. Management decision making could be more or less policy oriented and consistent, and union leadership could be more or less militant. The characterization of union leadership styles is in part arbitrary, and each individual exhibited some variation in degree of militancy apart from style. Not only was there a spectrum with respect to each major variable—task organization, union leadership, and managerial policy and organizational effectiveness—but there were numerous combinations of the basic characteristics of the major variables—environment, union, and management. A meaningful contrast may be made, for example, between the somewhat overdrawn description of the bricklayer department at Company 1 and a second high-rate department, the Company 5 coating department.

The coating department had a difficult union leadership situation. The committeeman was described as a strong advocate. Although his aggressiveness was not combined with a high degree of militancy at the third step (the chairman of the grievance committee was a radical, but was isolated within the local), thus presenting a less troublesome situation than in the bricklayer department, this committeeman nevertheless created a grievance rate that was the highest average rate of any department in the study. The task environment was a difficult one, with workers very responsible for quality of product and required to pay careful attention to work under close supervision. There were also frequent routine changes in product, though the department was free from technological change and from civil rights problems. If the departments were ranked in terms of degree of managerial difficulty, the coating department had a less difficult union leadership problem than the bricklayer department (particularly because of the absence of civil rights problems), and probably had a modestly less difficult task environment, though this might be debated; but the major difference was in the management variable.

Management in the coating department maintained very adequate control of quality and quantity of product and operating conditions. Although the department had incentive problems, the administration of the system was of high quality, creating few inequities, and there was an absence of pressure tactics. Maintaining a policy of not yielding to pressure tactics was facilitated by an absence of technological change. Though management did not use consultation—and such a policy would probably have been unworkable in that union situation—they did use a strong, systematic pattern of discipline which was an effective instrument of control even though it created many grievance challenges. But, in addition, or as a phase of a high degree of policy orientation, line management behaved as strong problem solvers. They also had available and utilized staff advice at the lower levels of the grievance procedure. Clearly, then, this department showing the highest statistical grievance rates found in the study and having a difficult task environment and union leadership problem was in no sense out of control from a management point of view. The high grievance rate reflected, among other variables, high managerial control through aggressive management, particularly with respect to discipline.

Again the relative grievance rates do not reflect the differences in the quality of the union-management relationships at Company 3. Here the low-grievance-rate department had a higher grievance rate than the high-rate department did in 1964 and 1966. Only in 1967 did the high department have a significantly relatively higher rate. But the grievance rate in the low department was raised by early technological change and continuing demotion problems in a particular zone seniority system. Yet the low department had a very good incentive operation, used consultation with the union, and demonstrated a union-management relationship of accommodation. The high department had a difficult union leadership problem and a long history of incentive conflict. Both departments exercised systematic discipline. Systematic discipline and effective lower level staff advice prevailed in the high-rate department, but it constituted a more difficult managerial problem than the high-rate coating department at Company 5. The statistical similarities in rates in the two structural mills do not portray the significant differences in the quality of the resolution processes.

Although each high-rate department had characteristics distinguishing it from its paired low-rate department, each high-rate department was very much an individual situation with its own particular combination of variables. In terms of degree of managerial difficulty the high-rate departments, though each exhibited some of the characteristics of the high-rate syndrome, ranged from the extremely difficult bricklayer department to the high-rate maintenance department in Company 4, which was only modestly different from the low-rate silicon department especially when consideration is given to the use of pressure tactics in the low-rate department, a form of conflict which went unrecorded in the relative grievance rates.

There is also, of course, a low-rate grievance syndrome which will only be described in broad outline. This syndrome starts with a favorable task environment which is both relatively stable (creating only routine problems) and largely free from technological disturbances. It has a low problem rate. The repetitive character of the problems facilitates the development and application of policy. The low-rate departments were all less complex to manage, in part because managerial policies were less difficult to develop and apply consistently. The coke ovens department illustrate how the typical managerial problems—equipment abuse, safety, and absenteeism—could be handled through programmed training and progressive discipline. Low problem rates and the routine character of the problems in association with inactive or problem-solving union leadership gave rise to low challenge rates and even lower grievance rates through informal resolution. Informal resolution was encouraged by the task environment, by the union leadership style, and by problem-solving managerial leadership acting within a policy framework. Even policy application could have its elements of flexibility, with the parties considering each grievance on its merits, since this type of resolution was compatible with the leadership relationships. One of the most interesting policy findings was the fairly typical use of management consultation with the union, a particularly effective reinforcing instrument of accommodation.

Again, however, the low-rate departments had their unique characteristics. The simplest department, the utilities department, did not use incentives, had a very low degree of close supervision, and was very routine in its employee and managerial characteristics. The coke ovens, though notable for its extremely unfavorable working conditions, was a low-pressure, low-problem department utilizing programmed management. The structural mill department, the steel-making plant, and the silicon department all had favorable task environments compared to their paired departments, but they were more complex than the utilities and coke ovens departments. Again the reinforcing relationships were task environment, problem-solving union leadership, effective managerial decision making through the utilization of lower level staff, an appropriate policy framework, and problem-solving line management leadership. The interactions created low-level, informal, rational problem resolution in association with low grievance rates.

The Implications of the Study for Management

The most obvious implication of the study for management is that it is necessary to achieve a more conscious awareness of departmental differences in grievance activity and the resolution process as well as to recognize the consequences which flow from these differences. The differences arise from varied environmental and behavioral characteristics in the departments and from intended and unintended consequences of corporate and plant dissimilarities.

Departmental differences are not simply, nor most importantly, those concerned with grievance activity. Grievance rates are symptoms of various combinations of environmental, union, and managerial variables reported and analyzed in earlier chapters. These rates reflect not only different combinations of variables but also degrees of each variable. Two fairly comparable levels of grievance activity may arise from very different circumstances and have very different consequences from a management point of view. For example, the coating department in Company 5, the highest-rate department found in the study, did not reflect loss of operating control or managerial policy weaknesses, and the high rate was in part accounted for by strong managerial action—the considerable use of disciplinary penalties to maintain quality of product. In most of these respects it was in marked contrast to the bricklayer department in Company 1, which had a roughly comparable high rate.

Particular consequences for management and for academic analysis do not flow automatically from differences in grievance rates. Although in this study high grievance rates reflected some type and degree of unfavorable circumstances, and low rates some combination of favorable circumstances, it is possible to conceive, for example, of a seriously disturbed department having a low grievance rate simply because the parties ignored the formal procedure. It is also possible to imagine a low rate in which management had largely abdicated, leaving little or nothing for the union to grieve. The interpretation of the significance of any given grievance rate or any given grievance rate difference requires analysis in depth and not a glance at the figures.

In a sense the extremely simple research design—the study of paired high- and low-grievance departments in a small sample of plants—gave rise to an enormously complex study. The considerable array of substantial differences in environmental, union, and management variables, with each department a very special case study, made generalization inevitably somewhat speculative and certainly difficult. The approach used unquestionably presented some extreme differences for investigation. A study of a much larger number of departments no doubt would have revealed for each variable something resembling more of a central tendency, with many smaller variations rather than the more extreme and somewhat disorderly differences presented here. Also, a much neater and academically satisfying research design, for example, might have demonstrated statistically the relationship of different degrees of favorable and unfavorable task environments with grievance rates, thus providing a simple hypothesis, but it would not have exposed the far more complex reality of grievance differences with which management must contend.

This study in fact suggests that a very important labor relations staff function should be the continuous diagnostic one of analyzing the nature of, the reasons for, and the consequences of differences in grievance activity and the resolution process. Such an analysis should review not only environmental and union variables but should look closely at managerial behavior, both line and staff. It is evident that management, although obviously aware of the problem, took a far too simple view of the grievance differences. The diagnostic survey should look closely at departmental differences in the managerial task, differences in managerial strengths in relation to the task, and differences in the consequences of the functioning of the resolution process.

Differences in the difficulty of the departmental managerial task stem from environmental and union influences. Technological impacts should give careful consideration apart from normal operating differences. Socioeconomic variables should be recognized as related to causes that typically go beyond managerial resolution, particularly at the departmental level. The diagnostic study should concentrate on the analysis of managerial strengths and weaknesses in relation to differences in the managerial task as revealed by the environmental and union analysis.

Variations in the managerial response, at the corporate and at the departmental level, and with line and staff management, could be singled out as a second implication of the study. A clear finding was that a lower-level, problem-solving resolution process required staff participation and availability at those levels to avoid forced resolution at a higher step. Just how line and staff should share authority and responsibility at the lower steps was less clear and will be discussed further subsequently. Another clear finding was that line management must operate within an adequate policy framework to avoid reinforcing the difficulties inherent in an unfavorable union and task environment. Finally, if management leadership is either autocratic or in effect inactive, thus forcing the major burden of resolution on the staff, the staff organization will be unable to cope adequately with the complexities of its task.

Before discussing what action management may take to adjust to differences in the managerial task, a digression is appropriate to examine the consequences that result from variations in the managerial task and in managerial strengths. Two very different resolution processes result: a high-level and a low-level one. The high-level resolution process is necessarily formal; it is removed from the operating context, has a different set of actors, and is a semi-legal resolution process dealing with rights and obligations under the contract. This high-level resolution process, with arbitration as its keystone, may or may not operate effectively or amicably. This high-level resolution process should be recognized as a very significant social invention, allowing the incorporation of no-strike clauses in labor agreements through its courtlike system of justice. At the same time, this high-level resolution process, though related to the low-level resolution process, that tended to prevail in low-rate departments, differs in important respects.

The low-level resolution process is more directly a problem-solving process carried out by lower-level supervision, lower-level union officials and employees which takes place on the factory floor, thereby allowing problems to be dealt with in their operating context and in a reasonably expeditious manner. The low-level resolution process is the essence of what a grievance procedure is designed to achieve, even though it, too, deals in rights and obligations under the contract, albeit in a broader and more flexible context.

There are substantial advantages to achieving a high degree of low-level resolution: (1) the effectiveness of lower-level supervision is enhanced, (2) an expeditious meeting of minds is achieved through settlement rather than through a delayed "decision" handed down from a higher level, and (3) problems are resolved as real operating problems, and not as legal and logical arguments over rights and obligations at higher and more removed levels of the grievance procedure. Achieving the highest possible degree of low-level resolution should be an important corporate-labor-relations objective.

The delays inherent in high-level problem resolution, and particularly in arbitration, were recognized in the steel industry some time after this study was completed. The result was the development of an informal, nonprecedent-setting, expedited arbitration procedure to supplement the regular arbitration process. By all accounts this new process has been successful, but it would not appear to meet the advantages of true low-level resolution listed above.

Recognizing differences in the influence of environmental variables and in union behavior, what are the action implications for management toward achieving lower grievance rates and an improved resolution process? A first action implication is to accept that the managerial task is much more difficult both in its environmental and union dimensions in some departments and that these differences must be analyzed, understood, and to some extent lived with. Neither line nor staff management should be judged as highly competent or a failure solely by reviewing grievance rates. In an easy-to-manage department an "average" formal grievance rate may be indicative of very mediocre managerial performance, while a fairly high grievance rate in a difficult department may be associated with superior managerial effectiveness. Although differences in grievance rates should be accepted as normal, high rates and a corresponding degree of high-level resolution should not be accepted fatalistically or on the basis of simplistic analysis. The objective of reasonably low formal grievance rates and, more important, a high degree of low-level resolution should be a key labor relations objective.

A second action implication for management is the choice of the type of shared line-staff authority it desires to implement as corporate policy. Making staff advice readily available to lower-level line management appears clearly advisable, but staff dominance seems to increase the risk of less effective operational and labor relations efficiency. Line management personnel require an effective voice in decisions bearing upon worker efficiency, even though they may not need to become specialists in rights and obligations under the labor agreement. There may well be no more than minor philosophical differences of opinion among various corporate managements at this point, but auditing both staff and line behavior must truly be a shared responsibility of top-level corporate line and staff. Certainly some of the companies studied achieved a better operational balance in sharing authority than did others. The degree of staff involvement at Company 4 should be regarded as a special case, with few applications to larger plants differently situated. However, the study demonstrates clearly only that staff advice must be readily available at lower levels to achieve low-level resolution. No definitive conclusions can be reached about what constitutes the best balanced form and degree of consultation; however, a most promising suggestion seems to be a "when requested" form roughly comparable to the practice at Company 5.

In a sense a more interesting third action implication is the adaptation of line and staff organizational structure and behavior to environmental and union differences affecting the difficulty of the departmental task. Some environmental task organizations made it extremely arduous for line management to function effectively in the grievance-resolution process. It should be possible in these environments to devise special operating schedules, or perhaps to deepen line management's organizational structure, or to create line management specialists, or finally, to invent other devices to prevent the line's organizational structure in a given environment from forcing a higher level resolution process. A less effective alternative might be special lower level staff arrangements to strengthen the lower level resolution process in difficult departments. Stereotyped organizational structures should not be allowed to handicap lower level resolution. As a special part of this adaptation, the organizational status of foremen could definitely be improved so this organizational asset would not be wasted to the extent that it appears to be at present.

Finally, special attention should be given to high-rate, difficult-to-manage departments. In this effort, shorter and longer term objectives might well be formulated. Nothing dramatic can be advocated for shorter term objectives. Substantive policies need to be clearly developed and consistently applied. This in turn may require new line leadership or special training for existing line management. Some modification in line organizational structure may be necessary to give greater opportunity for lower level resolution in difficult environmental settings. All of these suggestions may require specialized staff assistance. Through staff and line cooperation very careful first-step grievance answers need to be given to demonstrate effectively experience that appeal does not pay and that management will in fact be more flexible at the first rather than at later steps in the grievance procedure. There is a most significant difference between a high-rate department in which management has retained substantial operational control and one in which ineffective or inferior managerial decision making is interacting with a difficult environmental situation coupled with aggressive union leadership.

For longer range objectives in high-rate departments (assuming management has accomplished its shorter term objectives reasonably well), efforts should be made to begin an attitude-structuring program that would encourage communication between employees and management largely creating a more active role for foremen. Various personnel programs might be tried in an effort to improve employee morale. Dramatic changes can and have been made in high-rate departments by modifications in managerial leadership, organization, policy, and practice that have frequently been associated with subsequent changes in union leadership. Such improvements have been accomplished in entire plants by more positive employee relations policies and they are possible, even though difficult, in unfavorable high-rate departments. Intelligent action programs can be designed in the light of continuing diagnostic study.

The grievance-resolution process, formal and informal, is enormously complex. It is hoped that the analytical framework developed in connection with this steel study and illustrated by the chapter analyses has contributed new insights into variations in departmental grievance activity and the resolution process. A rational analysis of departmental grievance activity and the resolution process can suggest ways and means to build more positive employee relations policies and attitudes.

Bibliography

Crane, Bertram R., and Hoffman, Roger M. *Successful Handling of Labor Grievances.* New York: Central Book Company, 1956.

Derber, Milton, Chalmers, W. E., and Edelman, Milton T. *Plant Union-Management Relations: From Practice to Theory.* Urbana, Ill.: University of Illinois, Institute of Labor and Industrial Relations, 1965.

Derber, Milton, Chalmers, W. Ellison, and Stagner, Ross. *The Local Union-Management Relationship.* Urbana, Ill.: University of Illinois, Institute of Labor and Industrial Relations, 1960.

Dunlop, John T. *Industrial Relations Systems.* New York: Henry Holt and Company, 1958.

Fisher, Ben. "Arbitration: The Steel Industry Experiment," *Monthly Labor Review,* November, 1972, 7–10.

Fleishman, Edwin A., and Harris, Edwin F. "Patterns of Leadership Behavior Related to Employee Grievances and Turnover." *Personnel Psychology,* 15 (Spring 1962), 43–56.

The Grievance Process: Proceedings of a Conference, March 23–24, 1956. East Lansing, Mich.: Michigan State University, Labor and Industrial Relations Center, 1956.

Harbison, Frederick H. *Collective Bargaining in the Steel Industry: 1937.* Princeton, N.J.: Princeton University, Industrial Relations Section, Department of Economics and Social Institutions, 1937.

Kennedy, Van D. "Grievance Negotiation," in *Industrial Conflict*, edited by Arthur Kornhauser, Robert Dubin, and Arthur M. Ross. New York: McGraw-Hill Book Company, 1954.

Kuhn, James W. *Bargaining in Grievance Settlement*. New York: Columbia University Press, 1961.

——— "The Grievance Process," in *Frontiers of Collective Bargaining*, edited by John T. Dunlop and Neil W. Chamberlain. New York: Harper & Row, 1967.

Livernash, E. Robert. *Collective Bargaining in the Basic Steel Industry*. Washington, D.C.: U.S. Department of Labor, 1961.

Pettefer, J. C. "Effective Grievance Administration," *California Management Review*, 13 (Winter 1970), 12–18.

Phelps, Orme W. *Discipline and Discharge in the Unionized Firm*. Berkeley and Los Angeles: University of California Press, 1959.

Ronan, W. W. "Work Group Attributes and Grievance Activity," *Journal of Applied Psychology*, 47 (February 1963), 38–41.

Sayles, Leonard R. *Behavior of Industrial Work Groups*. New York: John Wiley & Sons, 1958.

Sayles, Leonard R., and Strauss, George. *The Local Union*. Revised edition. New York: Harcourt, Brace and World, 1967.

Schlicter, Sumner H., Healy, James J., and Livernash, E. Robert. *The Impact of Collective Bargaining on Management*. Washington, D.C.: The Brookings Institution, 1960.

Selekman, Benjamin M. "Handling Shop Grievances," *Harvard Business Review*, 23 (Summer 1945), 469–483.

Selby, Rose T., and Cunningham, Maurice L. "Grievance Procedures in Major Contracts," *Monthly Labor Review*, 87 (October 1964), 1125–1130.

——— "Processing of Grievances," *Monthly Labor Review*, 87 (November 1964), 1269–1272.

Somers, Gerald G. *Grievance Settlement in Coal Mining*. Morgantown, West Virginia: West Virginia University, Bureau of Business Research, College of Commerce, 1956.

"Steel Mill Foreman: The Job Nobody Wants," *The Magazine of Metals Producing*, July 1968.

Stieber, Jack W. *The Steel Industry Wage Structure: A Study of the Joint Union-Management Job Evaluation Program in the Basic Steel Industry.* Cambridge, Mass.: Harvard University Press, 1959.

Sulkin, H. A. "Comparison of Grievants with Nongrievants in a Heavy Machinery Company," *Personnel Psychology*, 20 (Summer 1967), 111–119.

Sweeney, Vincent D. *The United Steelworkers of America: Twenty Years Later, 1936–1956.* N.p., 1956.

Tilove, Robert. *Collective Bargaining in the Steel Industry.* Industrywide Collective Bargaining Series. Philadelphia: University of Pennsylvania Press, 1948.

Ulman, Lloyd. *The Government of the Steelworkers' Union.* Trade Union Monograph Series. New York: John Wiley & Sons, 1962.

U.S. Department of Labor, Bureau of Labor Statistics. *Grievance Procedures.* Major Collective Bargaining Agreements. Bulletin No. 1425-1. Washington, D.C.: Government Printing Office, 1964.

United Steelworkers of America. *Steelworkers' Handbook on Arbitration Decisions.* Pittsburgh: United Steelworkers of America, 1960.

Walton, Richard E. *Legal Justice, Power Bargaining, and Social Science Intervention: Mechanisms for Settling Disputes.* Institute Paper Series. No. 194. Lafayette, Ind.: Purdue University, Institute for Research in the Behavioral, Economic, and Management Sciences, Herman C. Krannert Graduate School of Industrial Administration, 1968.